86-1441

Springer Series in Physical Environment 15

Managing Editor
D. Barsch, Heidelberg

Editors
I. Douglas, Manchester · F. Joly, Paris
M. Marcus, Tempe · B. Messerli, Bern

Advisory Board
F. Ahnert, Aachen · V.R. Baker, Tucson · R.G. Barry, Boulder
H. Bremer, Köln · D. Brunsden, London · R. Coque, Paris
Y. Dewolf, Paris · P. Fogelberg, Helsinki · O. Fränzle, Kiel
I. Gams, Ljubljana · A. Godard, Meudon · A. Guilcher, Brest
H. Hagedorn, Würzburg · J. Ives, Boulder · S. Kozarski, Poznań
H. Leser, Basel · J.R. Mather, Newark · J. Nicod, Aix-en-Provence
A.R. Orme, Los Angeles · G. Østrem, Oslo · T.L. Péwé, Tempe
P. Rognon, Paris · A. Semmel, Frankfurt/Main · G. Stäblein, Bremen
H. Svensson, København · M.M. Sweeting, Oxford
R.W. Young, Wollongong

Springer
Berlin
Heidelberg
New York
Barcelona
Budapest
Hong Kong
London
Milan
Paris
Tokyo

Volumes already published

Vol. 1: Earth Surface Systems
R. Huggett

Vol. 2: Karst Hydrology
O. Bonacci

Vol. 3: Fluvial Processes in Dryland Rivers
W.L. Graf

Vol. 4: Processes in Karst Systems
Physics, Chemistry and Geology
W. Dreybrodt

Vol. 5: Loess in China
T. Liu

Vol. 6: System-Theoretical Modelling in Surface Water Hydrology
A. Lattermann

Vol. 7: River Morphology
J. Mangelsdorf, K. Scheurmann and F.H. Weiß

Vol. 8: Ice Composition and Glacier Dynamics
R.A. Souchez and R.D. Lorrain

Vol. 9: Desertification
Natural Background and Human Mismanagement
M. Mainguet

Vol. 10: Fertility of Soils
A Future for Farming in the West African Savannah
C. Pieri

Vol. 11: Sandstone Landforms
R. Young and A. Young

Vol. 12: Numerical Simulation of Canopy Flows
G. Groß

Vol. 13: Contaminants in Terrestrial Environments
O. Fränzle

Vol. 14: Saturated Flow and Soil Structure
H. Diestel

Vol. 15: Karst in China
Its Geomorphology and Environment
M.M. Sweeting

Marjorie M. Sweeting

Karst in China

Its Geomorphology and Environment

With 82 Figures and 36 Photographs

 Springer

Professor Dr. MARJORIE M. SWEETING (†)

ISSN 0937-3047
ISBN 3-540-58846-9 Springer-Verlag Berlin Heidelberg New York

Library of Congress Cataloging-in-Publication Data. Sweeting, Marjorie Mary. Karst in China: its geomorphology and environment / Marjorie M. Sweeting. p. cm. – (Springer series in physical environment; 15) Includes bibliographical references (p. 251–260) and index. ISBN 3-540-58846-9 (Berlin). – ISBN 0-387-58846-9 (New York) 1. Karst–China. I. Title. II. Series. GB600.4.C6S94 1995 551.4'47–dc20 95-3816

This work is subject to copyright. All rights are reserved, whether the whole or part of the material is concerned, specifically the rights of translation, reprinting, reuse of illustrations, recitation, broadcasting, reproduction on microfilm or in any other ways, and storage in data banks. Duplication of this publication or parts thereof is permitted only under the provisions of the German Copyright Law of September 9, 1965, in its current version, and permission for use must always be obtained from Springer-Verlag. Violations are liable for prosecution under the German Copyright Law.

© Springer-Verlag Berlin Heidelberg 1995
Printed in Germany

The use of general descriptive names, registered names, trademarks, etc. in this publication does not imply, even in the absence of a specific statement, that such names are exempt from the relevant protective laws and regulations and therefore free for general use.

Typesetting: Best-set Typesetter Ltd., Hong Kong

SPIN: 10480888   32/3130/SPS – 5 4 3 2 1 0 – Printed on acid-free paper

Preface

The writer has been occupied with karst problems for 50 years and first visited China in 1977 as leader of a Royal Society delegation of British geomorphologists – the first delegation from the UK to visit China since the inauguration of the new China and at the end of the Cultural Revolution (Sweeting 1978). It was clear from that visit that a study of the Chinese karst would help our understanding of the problems of karst development and that many of the geomorphological difficulties which at times had stultified karst thinking in Europe, might gain from the Chinese approach and ideas. First, for example, the Chinese karst is in its initial stages much more fluvial in origin than that of the Dinaric karst. Problems which worried European karst geomorphologists, such as dry valleys and the debates about karst base levels and the water table have not worried the Chinese so much. The development of the S Chinese karst has proceeded from an original fluvially eroded landscape which was later karstified. Secondly, the large area distribution of limestones in China, and the spectacular areas of karst, compared with the often relatively small areas of karst in Europe, have focussed our attention upon the fundamental issues of karstification, rather than upon the endless discussions on the origin of the small and less significant landforms in the karst.

Since 1977, the writer has visited China most years and has been able to work with many groups of Chinese colleagues in different parts of China. These include Yuan Daoxian and his colleagues in Guilin; Zhang Yingjun and Yang Mingde and their pupils at the Guizhou Normal University in Guiyang; Ren Mei E, Bao Haosheng and Wang Fubao and the geographers at Nanjing University; and Chen Zhiping and Song Linhua and colleagues from the Institute of Geography of the Academia Sinica in Beijing. To these and many other Chinese karst workers, the author is deeply grateful, and owes a great debt of thanks.

Stoddart wrote after the Royal Society visit in 1977, "Chinese geomorphology is essentially Chinese; it will be carried out by the Chinese and for the Chinese. They are dealing with their landforms and *their* resources, and in a sense geomorphology (especially perhaps of the karst) helps to establish a national identity in exactly the same way as the better-known achievements of palaeo-anthropology and archaeology. Western science has much to offer our Chinese colleagues but not to impose." (Stoddart, in Sweeting 1978). These comments are still true today, despite the exchanges which have taken place since 1977. The language barrier is in many ways the most important problem in any exchange. Many of the concepts which are familiar to Europeans and to N Americans are also indigenous to Chinese thinking; but many very competent Chinese karst geomorphologists are not able to express their ideas in a form which can be comprehensible in a European language. Most of the time we are communicating with those Chinese who have a grasp of a Western language. The same of course could be said of almost all Western geomorphologists in their attempts to explain their ideas to the Chinese. In general, Chinese thinking on landforms is much less analytical and more descriptive than in the West; it is also more pragmatic and related to the solving of the practical problems associated with irrigation and drainage in karst areas. This is illustrated by the early development of irrigation schemes and the long-term monitoring of karst springs. Data about springs and karst water go back to at least 453 B.C. (Yuan Daoxian 1981).

Although the karst landscapes of S China have been studied and also painted by the Chinese for generations, it was only at the beginning of this century that Western scientists and travellers became sufficiently aware of their existence to write about and describe the tower and cone karsts of S and SW China. The karst landscapes of S China are among the great classic landscapes of the world – to be compared with the Grand Canyon country, the Swiss Alps or the Great Barrier Reef. The South Chinese karst is possibly the best example of the effect and influence of landforms upon pictorial art; many Europeans believed the depiction of mountains in South China in the Guilin area to have been invented by the Chinese for artistic reasons but the modern traveller can now see that such landscapes exist in reality (Cotton 1948; Swann 1956). Joseph Needham says in his famous work on Science and Civilization in China, "Before I saw them I often imagined that such pinnacles were the inventions of painters to fit upright hanging scrolls, but in fact they are a real and remarkable element in Chinese scenery." (Needham 1954). The fame of the Guilin Scenery has made it a candidate for a world hesitage site.

The aim of this book is to discuss what we know about the origin of these landscapes, and to introduce them to Western geomorphologists. There is a large body of literature in Chinese on the Chinese karstlands, of which only a fraction is referred to in this book; where possible, the Chinese references which are given have

English, French or German summaries. The book also endeavours to explain Chinese karst geomorphological concepts to western readers; throughout, the *pinyin* system of transcription from the Chinese is used. The first part of the book gives a general discussion of the background to the study of karst geomorphology in China. This is followed by a consideration of different regional areas of karst, particularly those in S and W China – these are all regions where the writer has had the opportunity to work or to assess the geomorphological problems. Throughout the book, the author has tried to make comparisons with karst regions outside China in order to assess Chinese areas in a world context.

Oxford, November 1994   M.M. SWEETING

Acknowledgements

I am grateful to the Royal Society for support of my first visit to China and for its continued support over many years. In addition, some of the work discussed in this book has been supported by the Royal Geographical Society and by the Research Committee of the British Geomorphological Research Group. The award of a Leverhulme Emeritus Fellowship enabled me to spend some months travelling in China in 1988–89. I am grateful for the facilities offered to me by the School of Geography in Oxford and by St. Hugh's College, Oxford. Mrs. Ria Audley-Miller and Mrs. Kathleen Carne helped to type the original manuscript. Help in the transcription from the Chinese was given by Hu Mengyu. I am very grateful to Prof. Yuan Daoxian, Prof. Bao Haosheng and Prof. Zhang Zhigan for their careful reading and criticisms of the original text; it was also read in its early stages by Dr Peter Bull. I am also indebted to Prof. Ian Douglas and Prof. Dietrich Barsch for their suggestions in the later stages of the book. In China, I have received immense help from the Karst Institute (Ministry of Geology and Mineral Resources), Guilin; the Department of Geography, Guizhou Normal University, Guiyang; the Department of Geo and Ocean Sciences, Nanjing University; and from the Institutes of Geography, and of Geology, Academia Sinica, Beijing. I have spent many enjoyable years with my Chinese colleagues, travelling in, and discussing with them, some of the most spectacular landscapes in the world; this book is for them. This work is also part of the I.G.C.P. project 299.

Contents

1	**The Physical Context of Karst in China**	1
1.1	The General Physical Setting	1
1.2	The Distribution and Nature of Carbonate Rocks in China	11
1.3	Tectonic Controls	19
1.4	Present-Day Climatic Controls	21
1.5	The Importance of Quaternary Climatic Change	23
1.6	The Distinctive Character of the Chinese Karst Environment	31
2	**The Significance and History of Karst Studies in China**	32
2.1	Western Knowledge of Karst in China	32
2.2	The History of Development of Karst Studies in China	33
2.3	The Resources and Attitudes to Karst Science in China	41
3	**Karst Terminology and Karst Types in China**	42
3.1	Karst Terminology	42
3.2	Karst Types	45
3.3	The Main Types of Chinese Karst	49
	3.3.1 Peak Forest (Fenglin) Guangxi Type	49
	3.3.2 Plateau-Canyon Type (Guizhou Type)	50
	3.3.3 High and Medium Mountain Canyon Type	51
	3.3.4 Periglacial Type (Tibet Platcau Type)	53
	3.3.5 Hill and Wide-Valley Type (Central Shandong)	53
	3.3.6 Loess Mountain Valley Type (Shanxi Type)	54
	3.3.7 Buried Karst or Deep Karst	55
	3.3.7.1 The Bohai Gulf Type	56
	3.3.7.2 Sichuan Buried Karst Type	56

4	**The Guilin Karst**	58
4.1	Introduction	58
4.2	Peak Forest (Fenglin)	66
	4.2.1 Caves in the Fenglin	72
	4.2.2 Hydrology of the Fenglin	76
4.3	The Peak Cluster (Fengcong)	79
	4.3.1 Caves in the Peak Cluster	84
4.4	Age and Origin of the Guilin Karst	87

5	**The Cone Karsts of Guizhou**	93
5.1	The Cone Karst of Shuicheng	98
5.2	The Canyons of Guizhou	104
5.3	The Caves of Guizhou	113
5.4	The Slope Zone of Maolan	114
5.5	The Age and Origin of the Karst in Guizhou	117

6	**The Karsts of Yunnan**	120
6.1	Introduction	120
6.2	Conclusions on the Karst of Yunnan	136

7	**Karst in Other Parts of South China**	137
7.1	Karst in Southern and Western Guangxi	139
7.2	Karst in Hunan and Hubei and the Changjiang Gorges	143
7.3	Karst in the Lower Changjiang Region	148
7.4	Conclusions on the Karst of South China	150

8	**The Karsts of North China**	151
8.1	Introduction	151
8.2	Karst in the Qinling Mountains	154
8.3	The Shanxi Plateau	155
8.4	The Karst of Shandong	163
8.5	The Karst of the Zhoukoudian and Xishan Areas, SW of Beijing and Other Hilly Areas near Beijing	169
8.6	Karsts in NE China	176
8.7	Deep or Buried Block Mountain Karst in N China	178
8.8	Conclusions on the Karsts of N China	179

9	**High Altitude Karst: The High Mountain Karst of West Sichuan**	181
9.1	The High Mountain Karst of West Sichuan	181
10	**The Karst of Tibet and Other Parts of Chinese Central Asia**	196
10.1	Introduction	196
10.2	The General Context of Karst of Tibet	197
10.3	Chemical Features of Karst Groundwaters in Tibet	201
10.4	Surface Karst Features	205
10.5	Caves and Cave Sediments	212
10.6	Some Conclusions on the Tibetan Karst	218
10.7	The Salt Karst of the Qaidam Depression	220
10.8	Conclusions on the Karst in Tibet and Chinese Central Asia	221
11	**Karst Hydrogeology and Chemical Characteristics of the Karst Water**	223
11.1	General Comments and Behaviour of Karst Water	223
11.2	Contrasts Between Karst Water in North and South China	225
11.3	Karst Water Heterogeneity	232
11.4	Epikarstic Water	236
11.5	The Chemical Quality of the Karst Water	239
11.6	Conclusion	241
12	**The Position of China in World Karst Studies**	242
12.1	Geological Influences	242
12.2	Physiographical Influences	244
12.3	Climatic Influences	245
12.4	Influences of the Quaternary Period	247
12.5	Influences of Land Exploitation	248

References 251

Subject Index 261

1 The Physical Context of Karst in China

1.1 The General Physical Setting

The distribution of carbonate rocks on the earth's surface is uneven and irregular. Large areas of the world, like much of the shield of Africa have little or no carbonate deposits, but in China, carbonate rocks outcrop widely at the surface or exit at shallow depths below the surface, covering 1.3 million km^2 or one-seventh of the country's territory. About one-quarter of the world's total of terrestrial carbonate rocks occurs in China. China is an immense country; its area is almost 9.6 million km^2 and it comprises about 6.5% of the world's total land area. About 10% of China's rural population live in the karstified tropics of South China; 61% of the food supplies and 93% of the rice are produced there (Barbary et al. 1991). The major karst areas in China are located in densely populated and economically important regions, particularly in the northeast, south and southwest (Figs. 1 and 2). Moreover, the distribution of karstic rocks is so widespread that almost every geomorphological type of karst can be found in China (Ren Mei et al. 1982). If covered and buried karsts are included, in addition to surface karst, up to one-third of the country is involved. The management and development of the karst areas thus form important sectors of the economy. Almost one-quarter of the total water resources of China is of karstic origin (Yuan Daoxian et al. 1991), and a good proportion of the Chinese petroleum reserves occurs in karstic rocks. Limestones and karst in China thus permeate almost every aspect of its physical geography and no other country of comparable size (Russia, Brazil, USA or Australia for example) is so dominated in its modern geology and geography by karstic problems.

Geologically, China is situated almost entirely on the extensive Eurasian plate. In the southwest, this plate collides with the Indian and Burmese plate, while in the southeastern part its margin forms the subduction zone of the Philippine plate (Fig. 3). The collision of the Indian plate and the subduction of the Pacific plate have had very different tectonic effects. In the west is the high plateau of Tibet and the world's thickest continental crust, while in the east the subduction has resulted in the doming and convection of the mantle in the eastern part of China and in a change to thinner crust. To the north, within the Eurasian plate are the tectonic terranes of Siberia. Within China itself, there are three cratons of Precambrian rocks, which can be said to make up three microplates (Holland 1990). The differences in nature, "tectonic position

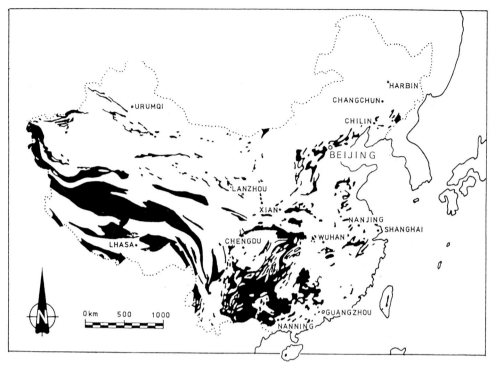

Fig. 1. Distribution of carbonate rocks in China

and mode and time-mutual adjustments and actions between the various blocks within China" have led to the formation of complex types of neotectonics (Gao Jiadong and Shao Shixiong, quoted in Meyerhoff et al. 1991). China is one of the most active tectonic regions in the world and the intraplate dynamics in this area pose problems for plate tectonic models (Holland 1990). This neotectonic activity is an important control in the development of the karst.

From the point of view of the nature and form of manifestations of the tectonic movements and history of evolution of the crustal surface, China can be divided into three megaregions – western China, eastern China and the Peri-Pacific megaregion (Gao Jiadong and Shao Shixiong, quoted in Meyerhoff et al. 1991). The Tibet plateau is part of the western megaregion (western China) and also part of a broader zone of deformation from the Himalayan thrust front to Lake Baikal. It has a dissected average elevation of 5000 m and is equal in area to the US Cordillera and the Rockies (Dewey et al. 1988). The Tibet plateau consists of great slices of terranes accreted to Eurasia; these terranes are generally oriented east–west (Fig. 4). The two most southern terranes, the Qiangtang and Lhasa terranes, came from Gondwanaland. Paleomagnetic data show that there were successively wide palaeotethyan oceans during the late Palaeozoic and early Mesozoic and a Neotethys which was at least 6000 km wide during the mid-Cretaceous

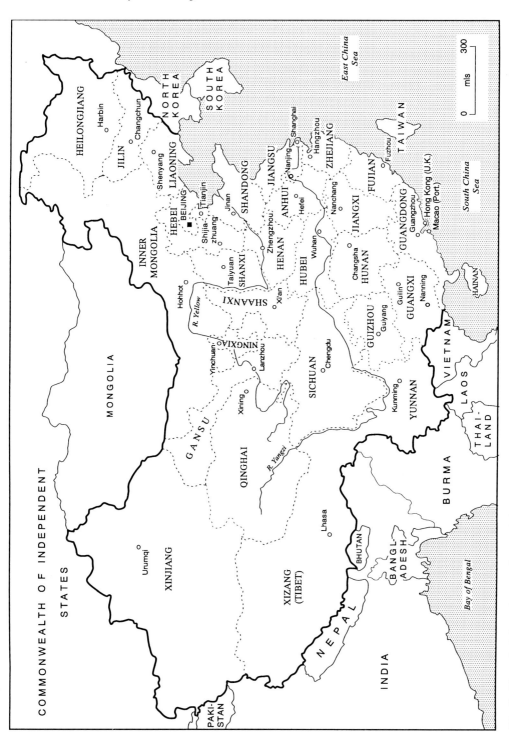

Fig. 2. The administrative provinces of the People's Republic of China

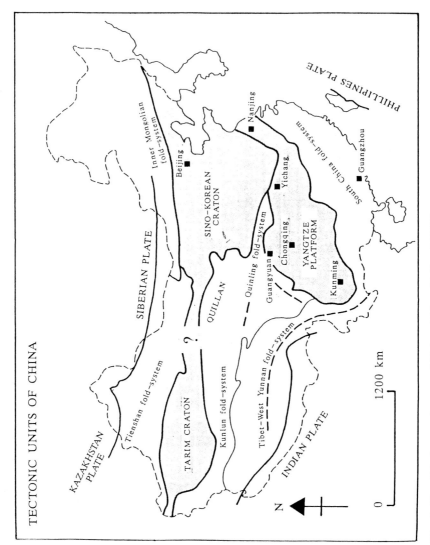

Fig. 3. Simplified tectonic map of China. (After Holland 1990)

The General Physical Setting 5

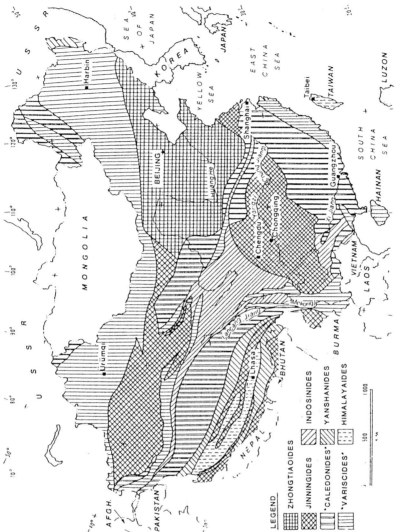

Fig. 4. Tectonic provinces of China. (After Meyerhoff 1991)

(Dewey et al. 1988). These Tethyan oceans account for the widespread developments of limestones on the Tibet plateau.

There is a great contrast in structural style between the Himalayas and Tibet. "The Himalayas have suffered some 80% shortening and are dominated by south-verging, gently northward dipping thrusts. By contrast, Tertiary structures in Tibet are thrust ramp basins containing folded red beds in a crust that has suffered a maximum 50% shortening" (Dewey et al. 1988). The thrust asymmetry was predetermined by the asymmetrical structure of the northern edge of India. Tibet has resulted from the bulk homogeneous shortening of the Asian lithosphere.

China is not only a large country, but it is also very mountainous – with mountains, hills and plateaux occupying 65% of the total area. The physical framework of the land of China has been clearly described by Zhao Songqiao (1986) and Zhang Zhigan (1980a). Geographers in China recognize traditionally two latitudinal lines and one longitudinal line of landscape division. These are (1) the Kunlun–Qingling–Dabie mountains, (2) the Yinshan–Yanshan mountains and (3) the Hengduan–Longmen–Liupan–Helan mountains (Figs. 3 and 5). These lines divide China into five major natural landscape zones: S China, N China, NE China, Qinghai-Xizang (Tibet) and NW China. The distribution of these five major landscape regions is closely related to the structure of the earth's crust in China. The five major natural landscape regions represent five blocks of the continental crust – the foundations of the natural landscape regions. Thus, China's landforms show a close correlation with geological structure. The Hengduan–Longmen–Liupan–Helan line is approximately 106°E longtitude and forms the boundary between the megaregions of East and West China; these two major parts of China have a quite different orientation of landforms. In western China, the landform units are arranged NW–SE or E–W, whereas in E China, they are arranged NE–SW or NNE–SSW (Fig. 4). In West China, there are many inland Basins (Tarim and Junggar for example) and block mountains (TianShan, Ordos plateau). However, E and S China are essentially made up of parallel NE–SW uplifted and down-warped belts and folds. Furthermore, on a relief basis, China is considered to be made up of four major relief steps; these are from west to east (1) the Tibetan–Qinghai plateau (over 4500 m); (2) the Yunnan–Guizhou plateau and the Shanxi plateau (1000–2000 m); (3) the East China plain (0–200 m) and (4) the present Continental Shelf area of the E and S China sea (Zhao 1986). The five major blocks and the relief steps control the present relief types in China and influence the distribution of karst types in the major landscape areas. The characteristics of the natural conditions of these major regions are given in Table 1; and the profile of the main relief steps is shown in Fig. 6.

Ren Mei, Yang Renzhang and Bao Haosheng in their book, *An outline of China's Physical Geography*, give a more complex physical division, in particular citing the Nanling mountains (or South Ranges) as being the divide between central and south China (Ren Mei et al. 1985). The Nanling mountains

The General Physical Setting

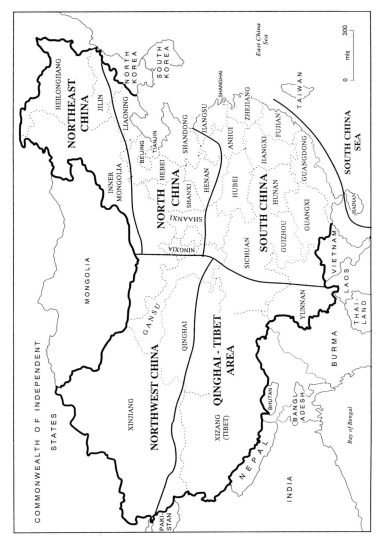

Fig. 5. The main physical regions of China

Table 1. Natural conditions and distribution of karst types in the six major landscape regions. (Zhang Zhigan 1980a)

Natural landscape region	Drainage area	Topographic features	Climatic zone	Crustal movements		Depositional period of main soluble rocks	Typical karst types
				Tectonic unit	Cenozoic tectonic movement		
South China	Chujiang (Pearl) and Changjiang (Yangtze) rivers	Low mountain – hill and medium mountain – plateau	Humid tropical, subtropical and temperate	Pre-Sinian para-platform and Caledonian fold system	Large-scale non-uniform slow uplift and moderate uplift; down-faulted and basin gradually dwindled; at their edges strong differential down-faulting movements occurred	Caledonian cycle (Lower Paleozoic), about 1.2 km thick; Variscan cycle (Upper Paleozoic), 2–5 km thick	Peak-forest type (Guangxi), canyon type (Guizhou), plateau – depression – basin type (Sichuan)
North China	Huang Huai-Hai rivers	Plain, low mountain – hill and medium mountain – plateau	Semi-arid temperate	Pre-Sinian paraplatform	Differential elevation and sub-sidence intense, thus forming down-faulted basins and massive uplift mountainous areas	All the cycles in the late Precambrian occur as troughs, 6 km thick; Caledonian cycle (Lower Paleozoic), 1.2 km thick	Loess mountain – wide valley type (Shanxi); buried block mountain (Gulf of the Pohal Sea)

The General Physical Setting 9

Northeast China	Heilong-jiang river	Plain, low mountain – hill and medium mountain	Semi-arid – humid temperate and frigid	Variscan fold system and Yanshanian fold system	Down-faulted basins gradually dwindled; large-scale non-uniform slow uplift and moderate uplift	Variscan cycle, occur sparsely	(Unclassified)
Qinghai-Xizang (Tibet)	High cold lakes	High mountain – plateau	High cold	Mainly Mesozoic fold system and Himalayan fold system	Large-scale rapid uplift since Pliocene times	Mainly Variscan cycle (1–2km) and Mesozoic cycle (3–6km)	Plateau relict forest-peak type
Northwest China	Inland	High mountain and basin	Arid-type high-mountain vertical zoning	Variscan and Caledonian fold systems with Pre-Sinlan paraplatform	Intense linear uplift zone with triangular or rhombic massive elevation and subsidence areas	Mainly Paleozoic deposits of geosynclinal type	High mountain – canyon type
Nanhai (South China Sea)	Tropical seas	Oceanic basin and islands	Humid tropical	Nanhai (South China Sea) platform (Variscan), partly transforming into oceanic crust	Large-scale subsidence, resulted in the Nanhai (South China Sea) basin	Cenozoic, Mesozoic and Upper Paleozoic	Modern coral reef-island

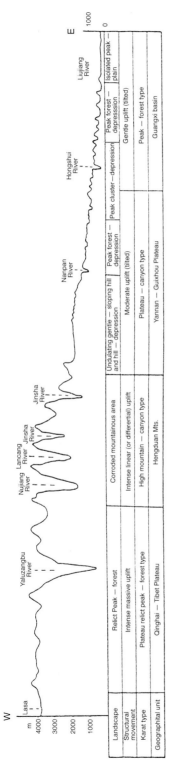

Fig. 6. Karst features in the region from Qinghai-Tibet plateau to the Guangxi basin, along the main relief steps. (Zhang Zhigan 1980a)

The General Physical Setting

are the chief geographical divide between central and South China, being 600 km from east to west and 200 km from north to south. They are, however, discontinuous and not as effective a barrier as the Qingling (Zhao Songqiao 1986).

Because China is part of the large land mass of Asia, it is drained by some of the greatest rivers in the world. The Chang jiang (Yangtze) and the Zhu jiang (Pearl river) in particular and their tributaries, bring down enormous volumes of water and have been able to cut deep canyons into the karstic terrain, up to 1000 m deep. The average annual flow of the Chang jiang reaches 31 055 m^3/s and ranks fourth among the great rivers of the world. The Zhu jiang's average annual flow is 11 075 m^3/s, about one-third that of the Chang jiang (Ren Mei et al. 1985; Table 2). Such rivers produce a dissection and rejuvenation of the karst, particularly of the underground hydrographic network and change the nature of the landforms. Nowhere else in the world is there such a combination of karst limestones and large actively dissecting rivers with such enormous discharges. The subtropical monsoon climate in this part of Asia is one of the highly contrasting seasons, which is reflected in the great variations in discharge of the great rivers. The physical geography of the Chinese rivers is thus an important aspect in the development of karst.

1.2 The Distribution and Nature of Carbonate Rocks in China

As indicated already, soluble rocks, limestones, are widespread (Fig. 1) particularly in the main southern and eastern regions, which form the basis of the Chinese karst. In addition, soluble evaporites cover large areas of the country, particularly in the dry areas of the north and west – for instance in Inner Mongolia, Qinghai province, the Xinjiang Uygur Autonomous region and in N Tibet. Gypsum beds are associated with the Ordovician limestones in N China and with Triassic and Sinian limestones and with dolomites in Sichuan. In the far NW, the areas are basically desert and the limestones occur in regions of very low rainfall – where the precipitation is not sufficient to develop karst in the normal sense at the present time.

The carbonate rocks of China occur in three main geotectonic units; the N China–Korean platform; the central and S China Yangtse platform; and the geosynclinalfold zone of S China (Fig. 3). The composition and age of the carbonate rocks in these three geotectonic units are shown in Fig. 7, taken from Zhang Shouyue and R. Maire in Barbary et al. (1991). The present distribution of limestones in China reflects (1) the distribution of ancient sedimentary basins and (2) the extent to which these rocks have been preserved since the Mesozoic. Since the Precambrian there have been four major carbonate and evaporite sedimentary basins developed in E China. Zhang Zhigan (1980a) estimates that soluble rocks account for up to 60% of the deposits in these basins. These four basins are as follows:

Table 2. Volume of flow of China's major rivers. (Ren Mei et al. 1985)

Name of river	Sea or lake they empty into	Drainage area (km²)	Length (km)	Average flow (m³/sec)	Total volume of flow (billion m³)	Depth run-off (mm)
Changjiang	East China Sea	1 807 199	6830	31 055	979.350	542
Zhujiang	South China Sea	452 616	2197	11 075	349.200	772
Heilong	Okhotsk Sea	1 620 170	3420	8600	270.900	167
Yarlungzangbo	Bay of Bengal	246 000	1940	3699	116.652	474
Lancang	South China Sea	164 799	1612	2354	74.248	412
Nujiang	Bay of Bengal	142 681	1540	2222	70.088	469
Minjiang	Taiwan Straits	60 924	577	1978	62.366	1024
Huanghe	Bohai Bay	752 443	5464	1820	57.450	76
Qiantang	East China Sea	54 349	494	1484	46.799	861
Huaihe	Huanghai Sea	185 700	1000	1113	35.097	189
Yalu	Huanghai Sea	62 630	773	1040	32.760	541
Hanjiang	South China Sea	34 314	325	942	29.710	866
Haihe	Bohai Bay	264 617	1090	717	22.600	85
Jiulong	Taiwan Straits	14 689	258	445	14.018	954
Yuanjiang	Beibu Bay	34 917	772	410	12.917	370
Ili	Balkhash Lake	56 700	375	374	11.790	208
Ertix	Kara Sea	50 862	442	348	10.985	216
Liaohe	Bohai Bay	164 104	1430	302	9.526	58

The General Physical Setting

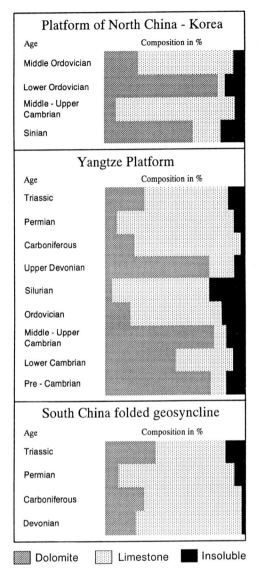

Fig. 7. The composition and age of carbonate rocks in various geotectonic units. (Barbary et al. 1991)

1. The late Precambrian Yao-Liao sedimentary basin. This is located in N Hebei and S Liaoning and the depositional centre lies in the Jixian area 100 km east of Beijing. Soluble rocks in this basin may be up to 6 km thick and at the transgression stage the area of the basin was approximately 200 000 km². The rocks are mainly dolomites with a high silica content.

2. The Lower Palaeozoic N China sedimentary basin (Cambrian to Ordovician). During this transgression stage, the whole of N China was more or less inundated – covering up to 720 000 km². The average depositional thickness of soluble rock is up to 0.95 km but may be as much as 5 km in places.

The rocks are a series of platform-type deposits in which lateral changes are very small. The lower deposits are mainly neritic limestones, the middle mainly neritic dolomites, and the upper include the famous Majiagou limestones, which consist of limestones and dolomites, these also contain evaporites and gypsum – which are important for the karst development in N China.

3. The Lower Palaeozoic Yangtse sedimentary basin (Sinian and Ordovician). This covers the Yangtse platform. At the maximum transgression stage this basin covered 600 000 km^2 and the average depositional thickness is 1 km. The facies zones are complicated and there are three major depositional environments – continental shelves and basins, basins of extensive seas, and platforms; the depositional environment of these limestones was more complex than those in N China.

4. The Upper Palaeozoic S China sedimentary basin (Devonian–Triassic). Most parts of S China were inundated during this period and over 1.25 million km^2 were covered by the transgression. As was deposited in Guizhou, the average depositional thickness of sediments is 1.2 km but in place 4–5 km. There were two major depositional cycles: (1) the Devonian – deposits about 1 km thick which occur in Hunan and Guangxi and 3–4 km thick in the Longmen mountains of Sichuan. (2) The Carboniferous-Triassic cycle, which is well represented in deposits on the Guizhou plateau.

It will be seen, as pointed out by Zhang Zhigan (1980a), that these sedimentary basins from the Late Precambrian to the Palaeozoic show that (1) the depositional extent of the carbonate rocks has increased continuously and that their depositional centres have constantly migrated southwards. (2) Each depositional basin has also tended to migrate southwards over geological time; in Mesozoic and Cenozoic times the depositional centres of carbonate rocks migrated south as far as the Sea of Tethys and into the South China Sea area as well. In Qinghai and Tibet there was also extensive deposition of carbonate rocks during the Mesozoic. This rather simplified scheme can be shown to be more complex in the light of interpretation in terms of plate tectonics (Yuan Daoxian et al. 1991).

The Ordovician limestones and others north of the Qingling-Kunlun line were deposited on the Sino–Korean or N China platform – a very stable area in which crustal movement after the Palaeozoic era was not intense and is mainly in the form of eperiogenetic movements (the uplift and subsidence of large structural blocks). The area is characterized by gentle folds, and large-scale faults and fractures. The middle Proterozoic and Cambro-Ordovician carbonate rocks were laid down on this platform. They suffered denudation during the late Ordovician to the early Carboniferous. There is also widely occurring fossil karst in the form of collapse columns which are particularly associated with the gypsum deposits in the Ordovician.

In NE China the limestones in the late Precambrian (Yan-Liao) sedimentary series in N Hebei and S Liaoning were deposited in a basin. They have been much more affected by later crustal movements those on the Sino–

Korean platform area; the stratigraphy resembles that of N China, but the folded structures, and hence the karst, are much more like those in S China.

SE China, south of the Qingling line consists of limestones deposited on the Yangtse platform and the S China geosynclinal fold system (Fig. 3). In these areas carbonate sedimentation was in (1) Upper Sinian to Upper Ordovician mainly in SW China and (2) from mid-Devonian to middle Triassic – where the total thicknesses are 3000–10000 m. The eastern area of the Yangtse platform was less affected by the later crustal movements; the Mesozoic Yanshan movements gave a structural grain of NE–SW, which is quite apparent. The karst is associated with a basin and fold structure.

However, in the west and southwest of S China, the intensity of crustal movement was greatly increased; many closed anticlines and synclines of small area were formed and also fractured rock basins. There is, therefore, a contrast between the conditions in the east and west (very generally) in S China. The geological context of the area of karst on the Yangtse platform is different from that in SW China. Moreover, in the extreme eastern parts of the Yangtse platform, as in Fujian, many of the carbonate rocks were never deposited as this area formed a plate margin.

The karst areas in Tibet are associated with the Cainozoic and Quaternary uplifts in the Himalayas. Here, both Ordovician and Mesozoic limestones have suffered intense crustal and thrusting movements from the south as at Quomolongma (Everest). They also occur within the east–west bands of uplifted Mesozoic beds on the Tibetan plateau. In Tibet, active east–west tensional faults are cut by north–south tensional grabens, and intersect the limestones. The conditions are quite unlike those of SW China.

The Sinian and Palaeozoic rocks have been continuously subjected to deformation and destruction since the Mesozoic. The present-day burial conditions of the soluble rocks can be grouped into four types (Zhang Zhigan, 1980), which are denoted as *bare*, *covered*, *buried* and *preserved*. Since the Mesozoic, the eastern margins of Asia have been subject to movements of the Pacific plate and three zones of uplift and three zones of subsidence have developed, arranged alternately from east to west. In general, the state of preservation of the soluble rocks tends to improve from east to west, partly because of magnetic activity along the Pacific plate margin. In Guangdong, Fujian and Zhejiang which belong to the second uplift zone, the soluble rocks have also been largely removed by denudation. In the second subsidence zone in Guangxi, Hunan, Jiangxi, Anhui, Shandong, Hebei and Liaoning, 60% of the soluble rocks have been preserved (but are often buried). In the third zone of uplift, 40–60% of the soluble rocks have been preserved and this is the most widespread *exposed* karst in China. The loess plateau of N China is in this zone; here the soluble rocks are covered locally by loessic deposits, from 10–100 m thick. In the third subsidence zone, there are platform-type depressions, as in the Sichuan basin and in the Ordos basin. Here the primary blanket of rocks has not yet been swept away, so that the soluble rocks are well preserved though buried and the original geochemical environments remain. As a result *palaeokarst* is abundant. The most obvious palaeokarst is the Caledonian orogeny disconformity between Ordovician limestones, and the gap between the middle Carboniferous and the upper Permian in S China. Palaeokarst is also seen in the karstification before the Jurassic deposition in Sichuan and in the karstification before the deposition of the Tertiary beds (Zhang Shouyue 1989).

Palaeokarst and fossil karst is karst which has formed and developed mainly during past geological time (Table 3). In a broad sense palaeokarst is that developed before the Quaternary (Ren Mei 1982), though the term palaeokarst in East China is often applied in a narrow sense to that developed before the Cenozoic. Karstification in E China occurred in five periods: the Precambrian, Early Palaeozoic, Late Palaeozoic, Meszoic and Cenozoic periods – corresponding to tectonic cycles. The Cenozoic is the most important period of karst development in China. The most important early period of karstification in N China was between the Middle Ordovician to the early Carboniferous; in S China the early to late Permian and Middle to Late Cenozoic times were significant. The development of the karst in the past was influenced by the palaeoclimate which was controlled by the changing palaeolatitude as a result of tectonic plate mobility. The relics of the karst forms which originated in these older periods were altered by processes of denudation during the climatic changes in the Tertiary and Quaternary, and are now seen in the present-day scenery (Zhang Shouyue 1989); this is particularly true of the karsts in N China. In the Tarim (Talimu) basin, palaeokarst in Ordovician limestones is covered by 5000 m of later rocks.

Geologically China's karst can be divided into *bare karst, covered karst,* and *buried karst*. Karst features that occur in soluble rocks underlying non-soluble bedrock is known as *buried karst*. Hydrologic chemical systems in buried karst are usually semi-open or relatively closed. The major buried karst areas are in Sichuan basin and the N China plain. In Sichuan, Proterozoic to Triassic limestones are covered by up to some 1000's of metres of the Mesozoic red sandstones. Caves in the Permian or Triassic limestones are frequently encountered in oil and gas wells at depths of 1000–3000 m. In the N China plain, carbonate rocks from the Proterozoic-Mid-Ordovician are buried by thousands of metres of Cenozoic strata.

In *covered karst*, the karst features in the soluble rocks are covered by unconsolidated or loess sediments, usually 10–50 m thick. Karst covered by Eocene clays occurs widely in S China, particularly in Yunnan. Usually, the covered karst is an open hydrologic system. Loess-covered karst is also important in the loess areas of Shanxi.

Bare karst occurs where the soluble rocks are exposed and the karst features are subaerial, and is the most conspicuous type of karst in China. It is characteristic of the uplifted areas, from Tibet to the Yunnan–Guizhou plateau, where, as a result of the uplift, the hydrological situation is very dynamic. In N China, bare karst occurs in the Shanxi plateau, where dry valleys are well developed as are the recharge areas of the major karst springs. The springs

Table 3. Karst epochs in the eastern part of China

Age	Area							
	South China					North China		
	Eastern Yunnan	Central Guizhou	Western Guangxi	Western Hubei	Eastern Sichuan	Shanxi	Shandong	
First karst epoch	Pre-Eocene	Shishan epoch	Loushan epoch	Western Guangxi epoch	Western Hubei epoch	Basin epoch	Beitai epoch	Central Shandong epoch
Second karst epoch	Eocene–Pliocene	Zhaolu epoch	Mountain–basin epoch	Peakforest epoch	Mountain–plain epoch	Yanjinggou epoch	Tangxian epoch	Tangshan epoch
Third karst epoch	Pliocene–Quaternary	Jinshajiang epoch	Wujiang epoch	Hongshuihe epoch	Changjiang (Yangtze) Gorge epoch	jialingjiang epoch	Loess epoch	Linjiang epoch

emerge at the boundary between the limestone mountains and the plains or valleys. Bare karst also occurs in the lower parts of E China where lateral circulation of water is more important. As a result of long periods of lateral corrosion, many isolated residuals rise from plains, as in the Guilin Tower karst area. Underlying the Guilin plain is a well-developed karst aquifer and in parts of Guangdong and in the S Guangxi basin, up to 2000 m of limestone is involved.

The limestones of the karst in China are in many ways unlike those of several other major karst areas. This is due to their relative antiquity – the most important karst in China being in the limestones of the Palaeozoic era. In areas of the Caribbean and in parts of the Indonesian archipelago (Borneo, Sulawesi), the limestones are often of Cenozoic age and have not always been affected so much by the diagenesis and alteration that have affected the more ancient and hard compacted Chinese limestones. The Caribbean and Indonesian limestones are generally less compact, softer and weaker; they give rise to excellent karst landforms but these are less spectacular and less upstanding than the striking karst hills of S China. In some parts of Borneo and Sulawesi, however, dense, pure and uniform Tertiary limestones have been sufficiently changed and metamorphosed to give rise to some significant karst landforms – such as in the Mulu area (Sarawak, Borneo) and the Maros area (Sulawesi), which rival (although in smaller areas) those in China. In the European "classical" karst areas of former Jugoslavia, the limestones are mostly Jurassic and Cretacecous (occasionally Triassic) and though they have been subjected to orogenetic forces, they show nothing like the diagenesis of the limestones in China. The Chinese limestones resemble much more the Devonian and Carboniferous limestones of NW Europe and the eastern USA where karst limestones are important, but where they do not outcrop on such a scale as they do in China. In many ways, the karst limestones of the Carboniferous in western Ireland (one of the most karstified areas of western Europe) most closely resemble the Devonian and Carboniferous limestones of Guilin and Guizhou, particularly in their diagenesis.

The diagenetic characteristics of the Chinese limestones are, therefore, of fundamental importance in the development of the karst. Recrystallization, dolomitization, pressure solution, cementation and types of pore space all affect the corrosion rates, cavern formation and hillslope form. For instance, in the Guilin area, large-scale cavern formation and the occurrence of steep isolated towers are more or less restricted to the outcrops of sparry allochemical limestones in that area – the relief of the micritic rocks being quite different. Dolomites normally give rise to low rounded cone-like hills, due to corrosional dissociation and the rapid mechanical disintegration of dolomite (Weng Jintao 1987a). The facies of the limestones depends upon the environmental conditions of deposition – whether platform, reef or basinal facies. The facies also affect the type of limestones deposited, the thickness of the beds and influence the bedding planes, and affect the results of diagenesis. These factors are well illustrated in the Guilin area. In general, the facies of the

Chinese karst limestones have been compacted and altered as a result of platform activity – both in N and S China; this means that vertical variations are more important than horizontal variations. Geosyclinal facies occur in the limestones of northwest and West China, this is in those areas where the climatic conditions are arid or semi-arid and where the karst is far less developed.

The porosity of most limestones in China is less than 2% and the permeability is almost equal to zero. The porosity of most dolomites is less than 4% and the permeability of the dolomites is also higher. Weng Jingtao (1987a) has also determined the compressive and tensile strengths of the rocks in the Guilin area. In the different stages of the karst process, chemical solution and physical denudation play different roles. During the initial stages, chemical solution is the most important; but in the middle and later stages physical collapse and destruction may become the main factor affecting the development of the landforms (Weng Jingtao 1987a).

Intercalations of marly and shaley beds in the limestone sequence equally affect the development of the relief. In Guizhou the development of the cone-like relief in marly limestones is sometimes more rounded and subdued with more gentle slopes than in those areas of continuous limestone beds. This is illustrated by the Triassic limestones in the surroundings of Guiyang city, between it and the airport. Cherty bands can also be important and affect the nature of the surface weathering – this is seen in the weathering of the Maokou limestones (Permian) in the Stone Forest (Shiling) of Lonan, in Yunnan province.

Though the occurrence of soluble rocks (limestones, dolomites, evaporites, etc.) is a prerequisite for the development of karst, other important factors control the development of karst. The most significant of these are tectonic uplift and an adequate precipitation. Such major controls also include other regulators. The nature of the limestone lithology and its structure, the type of diagenesis of the limestones and the geological history of the limestones are all geological factors. The runoff intensity of the rainfall, its type and amount, and the nature of the vegetation are essentially related to climatic factors. The main controls of the development of karst are, therefore, geological – having an indirect bearing on the relief; and climatic, having a bearing on the amount and intensity of precipitation and hence on denudation and corrosion rates.

1.3 Tectonic Controls

There have been many tectonic episodes which have affected China, and each episode has its own characteristics. These movements have controlled the deposition and subsequent evolution of the soluble rocks.

North and South China are composed of the pre-Sinian, i.e., Precambrian platforms and the Caledonian fold system. Von Richthofen (1882, 1912) called all the unfossiliferous sedimentary rocks between the underlying Archaean and the overlying Ordovician the *Sinian* – thus, originally, the Sinian was Precambrian to early Palaeozoic. It would now be called Proterozoic, with ages 850–800 million years at the base to 600–750 million years at the top (Meyerhoff et al. 1991). From the late Precambrian to the Variscan (including the Triassic), the platform areas were subjected to oscillating movements and soluble rocks, several thousands of metres thick, were deposited upon them. In the Mesozoic era, primarily as a result of the motion of the Pacific plate, the Indosinian and Yanshanian orogenic movements took place. These movements gave rise to the tectonic framework in E China with its general grain of structures NE-SW and NNE-SSW. This Mesozoic framework was destroyed or reworked in the Cenozoic when the Himalayan movements with their predominantly east-west trends gave rise to large-scale uplift and differential fault activity. The present-day landscape was, therefore, remoulded during the Himalayan period though probably much of the Indosinian-Yanshanian grain still remains (Holland 1990). The main tectonic zones of China are shown in Fig. 4 (simplified from Meyerhoff et al. 1991; Holland 1990).

Zhang Zhigan (1980a) has divided the periods of karst development into three stages. The first is the *early* or ancient stage: the Sinian to before the Triassic; the second he calls the *recent* stage: Jurassic-Miocene; and the third is called the *modern* stage: late Miocene to Holocene, i.e. from about 7–4 million years B.P. The *early* stage includes the periods of ancient karst formation when the soluble rocks were deposited. Quite often these ancient limestones have been sealed off and preserved. Ancient karst is widespread and forms deep-seated strata favourable for aquifers or reservoirs. After the Indosinian and Yanshanian movements and also the early Himalayan movements the blanket of clastic rocks which covered the limestones was gradually stripped off, giving rise to a new phase of karst development – the *recent* stage, from the Mesozoic to the end of the Miocene. The *modern* stage corresponds to the intense uplift in the Tibet plateau area and marks the appearance of deeply dissected valleys (Miocene–Holocene). The uplift and subsidence of the earth's crust in the late Tertiary and Quaternary is an important factor governing the process of karst development, particularly in S and W China. Uplift and subsidence dominate the changes in the base level of the karst and determine the hydrographic network in the karst areas and this, combined with climate, affects the processes of denudation. The fact that the continental crust of China is at a stage of intense neotectonic disturbance (the present Himalayan movements) is important for the strong development and different types of karst in China; this statement is illustrated by the development of periglacial karst in Tibet at altitudes of 4000–5000 m. The distinction between early, recent and modern karst in Chinese papers is different from that commonly used in North American and English publications, and is explained in Bosák et al. (1989).

1.4 Present-Day Climatic Controls

China is situated within the influence of the SE Asian monsoon and precipitation decreases generally from SE to NW. Also, because of the great latitudinal span from 53 °N to about 10 °N and its varied relief, China encompasses the warm and cold arid, the cold temperate, warm temperate and tropical zones; there are, therefore, many different climatic types. Rainfall is the most important climatic factor controlling karst, as it is the main input to the system; factors associated with rainfall, such as its intensity and highly contrasting seasonality and the intensity of runoff and surface flow are also fundamental. Precipitation varies from over 3000 mm/year in the SE to less than 25 mm in southern Xinjiang province (Fig. 8). Variations in this general trend are caused by variations in the direction of the coastline and the orientation of the mountain ranges. The distribution of the climatic zones in East China corresponds to a certain extent with that of the river drainage areas; the middle and lower Zhujiang (Pearl) river drainage belongs to the humid tropical and subtropical area; the middle and lower parts of the Changjiang (Yangtse) belong to the humid and north subtropical and temperate area; and the middle and lower parts of the Huanghe (Yellow) river drainage form part of the semi-arid temperate area. In S China, the index of aridity is less than 0.75; on the Shanxi plateau in N China, the index is 1.5 and in N Tibet it is 4.

The effects of temperature upon the karst system are much less clear (Fig. 8). It affects soil and plant development and hence biogenic CO_2 production. It is usually assumed, despite the greater solubility of CO_2 in cooler climates, that this increased solution is offset by the greater CO_2 production by vegetation in tropical humid regions, and that carbonate denudation rates are higher in warmer than in cooler regions. In fact, Zhang Zhigan gives the ratio for the denudation rates of the carbonate rocks in the three major drainage basins from north to south as 1:3:8, i.e. the rate of carbonate denudation in the Zhujiang basin is eight times that of the Huanghe. Table 4 (taken from Zhang Zhigan) gives a comparison between climatic factors and the denudation rates of the carbonate rocks in the three major drainage basins of eastern China. This kind of comparison usually identifies the tower karst of the Guilin area and other parts of Guangxi as being of a tropical karst type, and the N China karst as a temperate karst. However, as will be seen later, the tower karst of Guilin is a result of many factors, tectonic, lithological, time, fluvial and karstic denudation, of which climate is only one. The mistake is to correlate corrosion rates and intensity with distinct landforms. The view that climate is the major determinant is repeated throughout the Chinese literature (Zhang 1980a); this view is being superseded by looking at karst feature complexes (Yuan Daoxian et al. 1991).

Discussing the rate of turnover of plant organic matter in China, Ren Mei et al. claim that the amount of fallen leaves and branches per year from tropical rain forest in S China is about three times that of forest in the temperate zone (Table 5). The CO_2 content of soil air is also two to three times

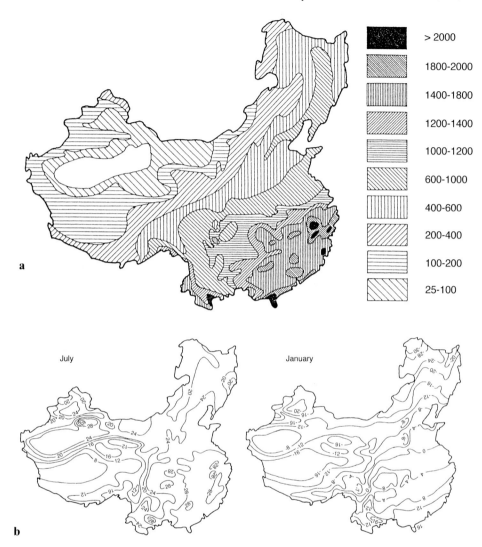

Fig. 8. a Annual average precipitation (mm). **b** Mean temperatures in July and January. (Barbary et al. 1991)

that in the subtropical soils compared with those in the temperate zone (Ren Mei et al. 1982).

These facts enabled Ren Mei to claim that the corrosion rate in limestones in central Guangxi (in tropical China) is 0.12–0.3 mm/year, whereas, in Hebei province (temperate N China), it is only 0.02–0.03 mm/year (see also Table 3).

Experiments with limestone tablets to assess limestone denudation rates in the different climatic zones of China (Fig. 9) show a good correlation between limestone denudation rates and precipitation, but it is not so clear with respect to temperature; the denudation rate depends more on precipitation than on latitude. Generally, the subsoil corrosion rate is much higher than the subaerial denudation rate, particularly in humid areas. A latitudinal comparison of points along 35°N latitude, from Japan (Akioshidai) to Golmud in the Qinghai province, also shows a good correlation with precipitation (Fig. 10). From these results, Yuan Daoxian concludes that the intensity of karstification (i.e. solution) is less in N China, where there is less than 500 mm rainfall, than in S China. In the more humid regions (which happen to be in the south), large cave systems and water circulation are more important and limestone denudation is much more intensive.

Temperature affects karst denudation processes in the high altitude permafrost areas of Tibet and the Himalayas. Snow and snowmelt may accelerate corrosional denudation in the Himalayas, but in the more arid parts of Tibet (at 4000–5000 m altitude) frost action is the dominant agent of denudation. The karst consists largely of frost derived fragments of limestone. Under such conditions, the denudation of the limestones is similar to that of other rock types; though there is some solution, corrosion rates are low and outweighed by the effects of frost action.

In general, therefore, the main climatic controls on karst are hydrological and related to precipitation. There is a delicate balance between surface runoff and the inputs into the karst system. The nature of the surface and underground relief depend upon rainfall and runoff variations, as well as the development of the initial routes of infiltration into the rocks (Smart 1988). The occurrence of dry valleys, steep initial slopes, and infiltration into the rock depend not only upon the geological factors, already outlined, but upon the nature, amount and type of the precipitation.

The interconnection between dissolution rates of the limestones and the structurally upwarped rates are combined in Fig. 11 from Lu Yaoru. Lu Yaoru (1980) also compared dissolution rates from areas in China with those in former Jugoslavia (Table 6). These dissolution rates have been calculated by means of the "tablet" method.

1.5 The Importance of Quaternary Climatic Change

Climatic change is particularly important because of the great inheritance factor in karst relief. The infiltration of water into the rock, together with the great purity of the limestones, means that there is little residual material left on the surface; furthermore, surface changes by means of solution are relatively slow, and erosion and corrosion are more active underground. Thus, as

Table 4. Corrosion rates of carbonate rocks of the three major drainage areas of eastern China. (Zhang Zhigan 1980a)

Drainage area	Climatic factors						Karst type	Denudation rate of carbonate rocks		
	Area (1000 km^2)	Rainfall (mm)	Yearly runoff (1000 m^3/s)	Run-off modulus (1/km^2)	Coefficient of comparison	Temp (°C)		Area name	Temp. (°C)	Denudation rate (mm/1000 years)
Zhujiang (Pearl) river	350	1000 to 3000	8.5	24	12	20 to 24	Peak forest area in the humid tropical zone, Guangxi	Luocheng, Guangxi	19	122.8
								Central Guangxi	20–22	120–300
								Tropical and subtropical humid area	20–24	40–120
Changjiang (Yangtze) river	1700	1000 to 2000	32.5	19	9.5	10 to 20	Plateau and mountain-canyon areas in the humid temperate and subtropical zones, Yunnan, Guizhou, Sichuan, Hunan and Hubei	Changjiang Gorges, Hubei	12–15	60
								Western Sichuan	9*	40–50
								Jingping Mountains, Mianning, western Sichuan	2–12	32–36
								Daba mountains, northern Sichuan	8–10	12.5–30

Huanghe (Yellow) river	730	200 to 800	1.5	2	1	6 to 10	Loess plateau-wide valley area in the semi-arid temperate zone, north China		10-30
							Warm temperate, subarid – subhumid area		20-30
							Northwestern Hebel	6–8	Denudation rate of carbonate rocks	9.5
							Middle section of the Taihang mountains (containing strata of gypsum and carbonate rocks)	6–10	Denudation rate of sulfats rocks	9.7
							Middle section of the Taihang mountains	6–10	8–10.5

* The author of this paper uses Corbel's method to calculate the results; in calculation, the $CaCO_3$ content is deducted from the rain-water.

Table 5. Morphological characteristics: amount of fallen leaves and branches in forests of different climatic zones in China

Type of forest	Locality	Amount of fallen leaves and branches (dry matter metric t/ha/year)
Tropical monsoon rain forest	Demenlong (Yunnan) 21°50′N	11.55
Subtropical evergreen broad-leaved forest	Huitong (Hunan) 27°N	4.53
Temperate and red pine forest	Xiao-Xinganling (Heilongjiang) 48°N	4.08

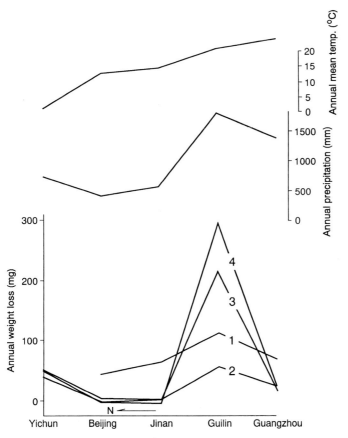

Fig. 9. Relationship between limestone denudation rate and climatic conditions in China; data from observation points in a north-south direction. *1* From subaerial limestone tablets (1.5 m above the surface); *2* from subaerial limestone tablets (on the surface); *3* from subsoil limestone tablets (15 cm below the surface); and *4* from subsoil limestone tablets (50 cm below the surface)

The Importance of Quaternary Climatic Change

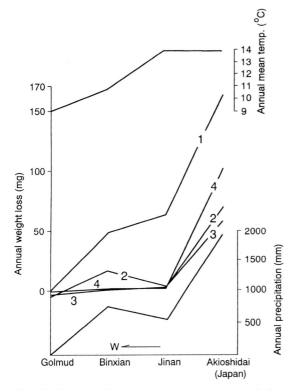

Fig. 10. Relationship between limestone denudation rate and climatic conditions in China; data from observation points distributed from east to west along 35°N lat. *1* From subaerial limestone tablets (1.5 m above the surface); *2* from subaerial limestone tablets (on the surface); *3* from subsoil limestone tablets (15 cm below the surface); and *4* from subsoil limestone tablets (50 cm below the surface)

Table 6. Dissolution rates from areas in China and in former Jugoslavia. (Lu Yaoru, no date)

	NW Heibei (China)	Middle Guangxi (China)
Dissolution rate per year (mm/y)	0.02–0.038	0.12–0.40
	Ljublijanica (former N Jugoslavia)	Neretva (former S Jugoslavia)
Dissolution rate per year (mm/y)	0.073	0.06

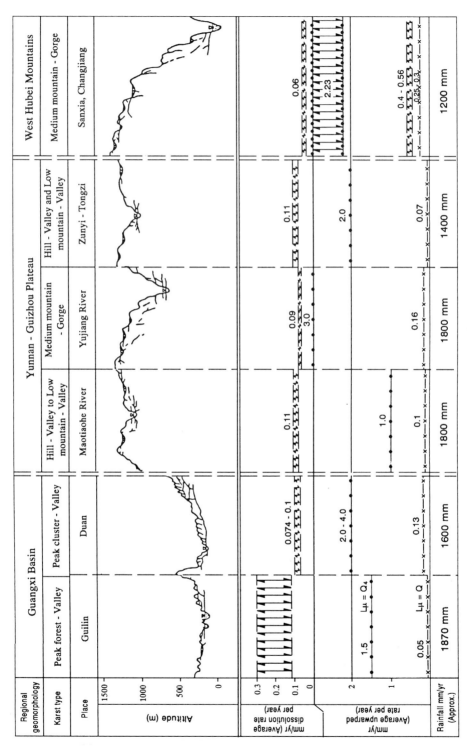

Fig. 11. Analysis of dissolution rates and structural upwarped rates in some typical karst regions of South and Central China. (Lu Yaoru 1980)

the French have said, the surface relief becomes *figé* or fixed and tends to form steady state landforms. Particularly in those areas where there is no frost or ice action, karst hills can remain as residuals in the landscape for a very long period of geological time. We now know that the fluvial deposits in the residual towers of the Guilin area are at least 900 000 years old and the caves and residual hills themselves are probably very much older. Thus, the persistence of limestone in the relief over geological time and the length of time the limestones have been exposed – is an important control in the development of karst. Because of the time factor, palaeoclimates and palaeokarst are of great importance and indispensable to an understanding of the processes that have moulded the modern karst landscapes.

It has already been shown that the most important periods of development of karst in China were the Cenozoic and the Quaternary (Zhang Shouyue 1989). Little is yet known of climatic changes in the Cenozoic though it is believed that the early Tertiary was warm and dry and the later Tertiary was warm and humid (Lin Jinrong 1992). In N China, the record of the Luochuan loess–palaeosol sequence in the loess plateau is of utmost importance for the study of the world's Quaternary (Liu Tungsheng 1988). The Luochuan sequence has developed in the past 2.4 m.y. and has been confirmed by magneto-stratigraphic studies. Other Quaternary geological events in N China (fossil man, the appearance and disappearance of ancient lakes, etc.) can be compared with the loess–palaeosol sequences. These sequences reflect the multicycle nature of climatic change with oscillations of dry/cold periods when the loess was deposited and warm/humid periods when fine-grained deposits formed or soils developed. These geological events recorded by the Quaternary strata in N China closely reflect global climatic change, as well as regional environmental change in the past 2.4 m.y. (Fig. 12).

Stalactite (speleothem) and travertine formation is likely to have taken place during the warm/humid phases. Uranium series-dating of speleothems for the past 350 000 years throughout eastern China has established epochs of non-deposition and epochs of greater deposition of speleothems. By attributing epochs of deposition to the interglacials and non-deposition to the glacials, interglacials are recognized (Zhang Shouyue 1989). These phases, like the palaeosols in the loess sequences, compare favourably with the isotopic marine record for foraminifera from core V28-239 (Shackleton and Opdyke 1973).

The Quaternary climates of southern and SW China have not yet been studied. Cave sediments and tufa deposits will eventually provide some of the data necessary to unravel the Quaternary climatic phases which affected the karst in S China. The present evidence suggests that the changes in the Quaternary climates in South China affected the development of the karst in quite a different way and to a much smaller extent than they did in N China (Shi Yafeng 1992).

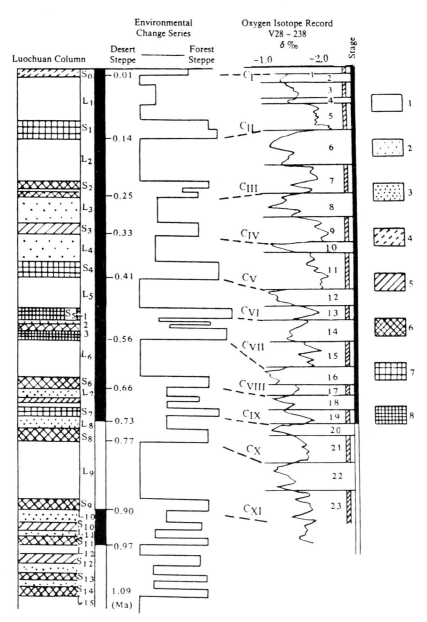

Fig. 12. Comparison between Luochuan natural environmental change sequence and V28–238 oxygen isotope records over the last 900 000 years. *1* Weakly weathered loess; *2* moderately weathered loess; *3* strongly weathered loess; *4* black loam; *5* carbonate cinnamon soil; *6* cinnamon soil; *7* leached cinnamon soil; *8* brown cinnamon soil. (Lui Tungsheng 1988)

1.6 The Distinctive Character of the Chinese Karst Environment

Though there is a great range of geology and climate in a country as large as China, some general points about the karst can be made. Except in Tibet, the karst limestones are of Palaeozoic age, crystalline and of great tensile strength. In South China, the limestones have been subjected to intense fluvial dissection by large rivers in a wet climate over a long period of geological time; in North China, despite some humid phases, the karst has evolved in generally drier, subhumid conditions. All areas of karst in China have been affected by differing types of Tertiary and Quaternary tectonic activity, including individual block movements and with compressional effects along plate margins. Many of these movements were contemporaneous with and contributed to Quaternary climatic changes. Thus the karst features of China have a long geological and multiphase history. This is illustrated by the discovery in some of the caves, not only of deposits over 1.6 million years old, but of possible deposits of Triassic age.

2 The Significance and History of Karst Studies in China

2.1 Western Knowledge of Karst in China

Ford and Williams (1989) have said that in Europe the "Mediterranean Basin is the cradle of Karstic Studies". In China, karst studies in general developed separately from the West. The relatively late discovery by outsiders of the Chinese karstlands was partly due to civil unrest and difficulties of travel in China. The Sino-Japanese war, followed by the revolution and then the "Cultural Revolution", meant that there were few opportunities for scientists to work in China. Travellers and archaeologists, like Sven Hedin and Aurel Stein, worked in what is now Chinese central Asia in the 19th and early 20th centuries (Hedin 1898; Stein 1912). The first geological reconnaissances in N China were made by von Richthofen from 1866–1872, paying particular attention to the loess landscapes (von Richthofen 1877–1912). However, little attention was given to South China. It was not until the 1920s and 1930s that Western scientists made any contribution to the study of the limestones in China. The first European description of the tower karst is believed to have been given by Handel-Mazetti, a botanist in 1926 and discussed by the geographer, O. Lehmann (Handel-Mazetti 1926; Lehmann 1926). In fact, the botanists and "plant-hunters" were some of the most important travellers in S and W China in the early years of this century (Sweinfurth and Sweinfurth-Marby 1975 on Kingdom Ward). Much significant work was done by French geologists, including the Jesuit, Teilhard de Chardin; de Chardin not only assisted in the excavation of the Peking Man site at Zhoukoudian, but also studied the Devonian limestone sequences of S China and made the first section across the tower karst limestones of the Guangxi basin near Guilin in the 1930s (Teilhard de Chardin et al. 1935). Bouillard also surveyed the caves of Yunshui in the Shangfang mountains near Peking in 1924; these caves are developed in siliceous dolomites of the Wumishan formation, middle Proterozoic (Bouillard 1924). The American geologist, Barbour (1930) discussed travertines in N China.

Tower karst similar to that in S China extends southwards into Vietnam and Laos – the old French colonial territory of Indo-China (Fig. 2). In this area, French geologists and geographers described the residual tower karsts of SE Asia in their work in the 1920s and 1930s (Cuisinier 1929). H. von Wissmann, the German Scholar, also travelled in SE Asia and included

S China in his observations on tower karst. The best-known article by von Wissmann appeared in "Erdkunde" in 1954, in a symposium held in Frankfurt on the influence of climate on karst landforms – and in which the climatic factors believed to influence karst development are discussed (von Wissmann 1954). European and American ignorance of karst in China until quite recent decades can be seen as late as 1972, when a book was published on *Karst: Important Karst Regions in the Northern Hemisphere*, in which the largest area of karst, not only in the northern hemisphere but in the world, was not even mentioned (Herak and Stringfield 1972).

2.2 The History of Development of Karst Studies in China

Karst phenomena have been known in China from very early times. Caves in N China and their hydrography were described in a book written before 221 B.C. A map drawn in 168 B.C. (Han Dynasty) shows tower karst in Hunan province, and is the first map of tower karst in the world (Fig. 13; Ren Mei 1984). China, south of the Yangtse (Chiangjiang), was settled later than the Middle Area (Yangtse – Huanghe region). However, an early attempt at diverting rivers from the well-watered south to the more arid north is seen in the construction of the Ling Qu canal, some 50 km northeast of Guilin, which was built in 214 B.C.; it was built in the Qin Dynasty when Qin Shihuang ruled, the first Chinese emperor who tried to unify S China. The aim of the Ling Qu canal was to divert water from the tributaries of the Pearl river to those of the Yangtse and to convey troops and military supplies. It was so well built that in its renovated form it is still in use today. The area around Guilin was, therefore, known to the Han settlers by the date of its construction and became an important crossroad town.

Guilin itself in 111 B.C. was a small village, Shi-an, until 265 A.D. when it became Shi-an County. Guilin was an early tourist centre and the caves of Guilin attracted tourists and the Seven Star cave (Qixing Yan) was already a tourist cave; an inscription was carved at the entrance of the cave in 590 A.D. This is now partly covered by tufa deposition, which gives us an idea of the rate of deposition since 590 (N. Wei/Sui Dynasty). Li Daoyuan, a geographer in the N. Wei Dynasty (386–534 A.D.), wrote about the Lijiang river, described the Seven Star hills (Qixing Shan) and the springs; he also explored some of the caves and writes about cave pearls. The famous Guilin tower karst hills with their peaks and waters are the subject of many poems originating at least since the year 760 A.D., and many famous poets have written about the area. The Chinese people say that the Guilin landscapes with their hills and rivers are "the most beautiful mountains and rivers than in any other place". Liu Zongyuan wrote a poem about stalagmites in 827 A.D. and in the Tang Dynasty in 899 A.D. a book described many of the picturesque sites and caves in the Guilin area, Stone inscriptions in caves are common all over China from the

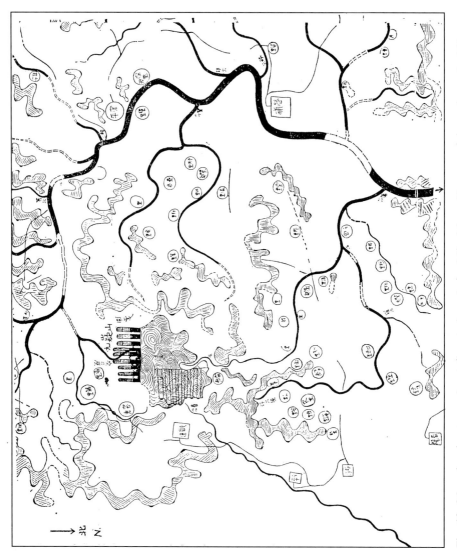

Fig. 13. Map showing tower karst (karst pinnacles) of Jiuyishan, Hunan province, drawn in 168 B.C. (ancient map from Han Tomb, Mawangte, Changsha)

Sui Dynasty (A.D. 581–618) right up to the Qing Dynasty (A.D. 1644–1912) (Ru Jinwen et al. 1991).

In North China, the great karst springs have been utilized also from early times. The Hongshan spring, S W of Taiyuan city in Shanxi province, has been developed since the Song Dynasty. According to a record on a stone carving, the spring was harnessed in 1040–1090 A.D. and divided into channels to irrigate 15 200 mu (1 ha = approx. 15 mu) of farmland (Yuan Daoxian 1981). People were also curious about the origin of karst phenomena. Thus, Fan Chengda in the Song Dynasty (1175 A.D.) gave as an explanation of stalactites (Shizhongru): "Milky water is dripping continuously in the cave, with the process of condensation taking place simultaneously" (Yuan Daoxian 1981); each stalactite "hollow in its centre is just like a goose feather" and "the milky bed as white as snow is formed by the condensation of stone solution". There were also discussions about the climate in caves. Jinan, the capital of Shandong province, is also known as "Spring City"(Yuan Daoxian 1981).

The first substantial scientific work on the karst and caves of S China is that of Xu Xiake (*Xu Hongzu*) (1587–1641 A.D. Ming Dynasty). He made an extensive journey in 1636–1641 A.D. to study the caves and geomorphology of the karsts of S and S W China where carbonate rocks occupy over 500 000 km². Xu Xiake was the first person in the world to discuss tower karst and to describe the different types of landscape in S China, particularly the tower karst (peak forest) of Guilin. He introduced the term f*englin* (peak forest), which is still used as a geomorphological term today. Xu Xiake travelled in the Guilin area in the summer of 1637 for 51 days and explored more than 100 caves. His book was very accurate for its time and he gives detailed accounts of the overflow caves that he explored, and of their chemical deposits.

Xu Xiake also travelled in many other parts of China, in journeys that lasted over 30 years. He was particularly interested in the determination of river valleys and their headwaters and his physiographic profiles of mountains and valleys must be among the first ever produced (Xu Hongxu 1642). The explorations of Xu Xiake are preserved in 16 volumes and include more than 690 000 Chinese characters. Among some of his observations, are those on the headwaters of the Yangtse river (Chang jiang). He also recorded the directions of cave entrances, cave sizes, and their different types of speleothems. A scientific meeting to celebrate Xu Xiake's work was held in 1990. Since Xu Xiake there have been further descriptions of karst phenomena, but until the late 19th and early 20th centuries, they did not constitute any real advances in the subject.

A recent assessment of Xu Xiake's work has been made by Hu Bangbo (1991). The *Diary of the Travels of Xu Xiake* (*Xu Xiake youji*), first published in 1642, has been now reprinted in classical Chinese by the Shanghai Guji Chubanshe (Shanghai Press on Ancient Books, reprinted 1982). A symposium to celebrate the 400th anniversary of his birth was held in Nanjing in 1986–1987 and a statue of Xu Xiake now stands outside the Institute of Karst Geology in Guilin (Photo 1). Needham (1959) stresses the importance of Xu

Photo 1. Statue of Xu Xiake outside the Karst Institute in Guilin

Xiake: "For more than 30 years, he perambulated the most obscure and wildest parts of the Empire, exposed to all kinds of difficulties and sufferings. ... His notes ... read more like these of a 20th century field surveyor than of a 17th century scholar. He had a wonderful power of analyzing topographical detail, and made systematic use of special terms which enlarged the ordinary nomenclature." (Quoted by Hu Bangbo 1991, and from Needham 1959, p. 524). The tracks of Xu Xiake in S China, from 1587–1641, are given in Ru Jinwen et al. (1991).

Chinese geomorphologists began to study the South China karst in a modern context during the Japanese occupation in the 1930s and 1940s, when the government was moved to SW China. These geomorphologists included Ren Mei and Yang Huairen. Also, two of the first Chinese modern scientists to look at the karst in a modern geomorphological context were Chen Shupeng (1965) and Zeng Shaoxuan. Chen Shupeng in particular was interested in the developmental sequences of the karst landscapes and was the first to call attention to the geomorphological zonation of the karst forms seen in Guangxi and Guizhou; he also realized the close connection between these zones and the karst hydrogeographical network. Chen Shupeng and others were also concerned about the extent of the Quaternary glaciation in S China. Prof. J.S. Lee (Li Siguang) had earlier published his ideas about the glaciation in S China – proposing that areas as far south as Guilin (25°15′N) had had extensive glaciers in the Pleistocene (Li Siguang 1939). These ideas generated many papers on the possible occurrences of till, glacial gravels and morainic deposits in the Guilin area, particularly those papers by Sun Dian Qing (1944) and Wang Kejun (1988).

Chen Shupeng, moreover, made many maps of caves and showed how accurate those made by Xu Xiake were. He was also interested in the different levels of caves and the terraces, and the levels of the hill summits and their fitting into geological time (Chen Shupeng 1957).

It was not until after the founding of the People's Republic of China in 1949 that active research into the Chinese karst by non-Chinese scientists became possible. After the liberation, scholars of the "Eastern block" in Europe were encouraged to visit the karst areas, in particular, Guilin. From the 1960s we have the work of Silar (1963, 1965), Gellert (1962), Kozarski (1962), and Balazs (1961, 1962). All these were concerned with the origin of the Guilin and N Vietnam tower karsts – regarded as the classic examples of tropical and subtropical karst development. These works contain some very perceptive ideas, particularly in relation to the developments of climatic geomorphology in Europe at that time.

Scientific work in China was restricted during the "Cultural Revolution", but after 1975 many overseas geologists and geomorphologists had not only visited the Chinese karst, but had been able to work on projects with their Chinese colleagues. These include the Sino-French hydrological project in the Guilin area (Yuan Daoxian and Drogue 1988); the Anglo-Chinese cave projects (Eavis and Waltham 1986); the work by P.W. Williams in Guizhou with the Guizhou Normal University Department of Geography (Williams 1978); and the work by the writer with the Karst Institute in Guilin and the Geography Department of Nanjing University (Sweeting 1978, 1990). In addition, the Chinese work itself has literally "exploded"; pent up for years by political and wartime difficulties and removed from Western ideas, much of the thinking and work on the karst areas of China has been brought out into the open and published, both in Chinese and English. The Ministry of Geology has established two institutes which have contributed in their different ways to the development of the knowledge karst. In N China at Zhengding (near Shijiazhuang in Hebei province) is the Institute for Engineering Geology and Hydrogeology founded in 1956. This institute, particularly its karst research division, has produced significant studies on N China on karst formation and hydrology. In S China the Institute for Karst Research at Guilin founded in 1976 has made great advances in karst geology and geomorphology, karst deposits, and problems of the karst environment. Other important work has come from karst groups in the Institutes of Geology and of Geography of the Academia Sinica in Beijing and from the geography and geology departments of the University of Nanjing, Guizhou Normal University, Chengdu College of Geology, and Guizhou Technological College.

Thus, since the founding of the People's Republic of China, in 1949, research into the karst has developed rapidly. The utilization of the karst, for water, for limestones, and for the metalliferous and non-ferrous materials associated with the karst led to enormous development in engineering and hydrogeology. Irrigation, hydroelectricity, industrial and railway construction all depend on detailed knowledge of the karst. Great numbers of people have collected data, and made topographic and geological surveys of the whole country. These data were collected by the famous geological teams, one of the best known in karst being the Liuzhou team led by Chen Wenjun. All-China Symposia on karst have been held in 1961, 1965, 1978 and in the 1980s. As a result of this activity, various books and maps have been published. An *Atlas of Hydrogeological Maps of the People's Republic of China* has been published together with a karst hydrogeological map of China which includes a karst hydrogeological map of karst areas in S China at a scale of 1:4 000 000. They show the general distribution of karst water in S China and its yield and its heterogeneity (Li Datong et al. 1985). Topographical and geological properties of different regions are also given. The Academia Sinica has produced a book on R*esearch of karst in China* (1979); it includes a map showing the distribution of karst in China on a scale of 1:10 000 000 and its regionalization (Zhang Shouyue 1979). Considerable advances have also been made in the preservation of cultural relics and the excavation of caves with archeological deposits, as at Zhenpi near Guilin and at the Peking Man's cave at Zhoukoudian, near Beijing. Numerous caves have also been surveyed and opened to tourism.

The Institute of Karst Geology was founded in 1976 at Guilin, but the building was not finished until 1981. It is now a part of the Chinese Ministry of Geology and Mineral Resources. There, all aspects of karst are studied, both from practical and academic points of view. It has encouraged the development of our knowledge of the karst, in particular, assisting many foreign visitors to work in the karst areas. There is also an adjacent Scientific Exchange Centre, a fine library and Museum of Karst. This institute has published important works including a journal – *Carsologica Sinica* which is of a high standard. In Guizhou, the geography department of Guizhou Normal University in Guiyang and the provincial authorities at Puding have also greatly encouraged foreign workers. The Nanjing University Department of Geography has also pioneered research into the karst areas and encouraged its postgraduates to do field work in parts of China as far away from Nanjing as Tibet, Sichuan, Guizhou and Zhejiang. Graduates from the karst studies group in Nanjing now pursue their work in many parts of China (Ren Mei pers. comm. 1992).

Though Beijing is far from the very extensive karst areas of China, the Institutes of Geology and of Geography of the Academia Sinica have established karst groups which have done notable work; some of these have been led by Chen Zhiping and Song Linhua and also by Zhang Shouyue. Members of these groups have contributed to not only important individual papers, but also to the symposia (Research of Karst in China) and collections of papers on karst. The Institute of Hydrogeology, also associated with the Ministry of Geology and Mineral Resources, is situated in Zhengding, near Shijia Zhuang in Hebei. Since so much of China is karstic, much of this work is concerned with karst hydrology. Much work has been done by members of this institute on the karst springs and karst collapses associated with the lower Palaeozoic limestones of N China. One member of this institute, Lu Yaoru, has produced a popular and well-illustrated volume on *Karst in China* (Lu Yaoru 1986).

Another aspect of Chinese interest in karst is their work in Tibet. The karst of Tibet has attracted attention, since the first work there during the Chinese scientific expeditions to Tibet in 1951–1953. Further expeditions took place in 1958–1961 and 1966–1968; mountaineering expeditions were organized in 1959–1960 to Everest (Qumolongma) and in 1964 to Xixabangma (Chui and Zheng 1976). A large team of scientists contributed to work during 1973–1976, after which a series of volumes were published by the Science Press of China incorporating the results of Tibetan Exploration Team of the Chinese Academy. Among these volumes have been: the Report of Science Expedition in the Everest Area (Chui and Zheng 1976); Studies on the Period, Amplitude and Type of Uplift of the Tibetan-Quinghai Plateau (Lin and Wu 1981); Tibetan Geomorphology (Chui 1983); and Geological and Ecological studies of the Qinghai-Xizang (Tibet) plateau (Academia Sinica 1981). Chui, in particular, was of the view that the karst landforms on the Tibetan plateau were relict palaeokarst forms developed under more humid tropical conditions in the mid-Tertiary, and that the karst landforms of the plateau could be compared with present-day tropical and subtropical forms in S China (Sweeting 1990). Most of these works are in Chinese with good English summaries.

The focus of karst studies in China is, however, inevitably in Guilin, in the Institute for Karst Geology. Guilin is favoured by its natural situation on the river terraces at the entrance to the Lijiang gorge and on the tower karst plain with its Devonian limestone peaks. Not only are members of this institute active in their research on karst problems, but Guilin is one of the most popular tourist destinations in China, attracting more tourists per year than all other areas except Beijing, Guangzhou, Shanghai and Xi'an. Thus, our European knowledge and literature of the Guilin tower karst is much greater than our knowledge of other karst areas in China. The 21st Congress of the International Association of Hydrologists was held in Guilin in October 1988. As a result of the untiring energy of Professor Yuan Daoxian, the institute's former director, the first UNESCO/IUGS project on karst is based in Guilin; this is the IGCP project 299 on geology, climate, hydrology and karst formation. From this project it is hoped that great advances on the formerly disparate studies in karst will be made. One of the first projects for the IGCP has been the first book in English on the karst of China: it is edited by Prof. Yuan Daoxian and contributed to by his colleagues in Guilin (1991). This book indicates to the world the significance and depth of thinking on karst in China and the problems that have arisen from its exploitation.

In 1993, the International Speleological Union held its 11th Congress in China and the Karst Institute of Guilin was one of the important centres of speleological activity. Outdoor leisure pursuits like walking and caving, unlike in Europe, are not important in China, However, recently a-China caving association has been set up under the Presidency of Professor Zhu Xuewen of the Guilin Institute, having members from many parts of China and meets about once a year.

2.3 The Resources and Attitudes to Karst Science in China

Yan Rong (in Chinese 岩溶) means rock dissolution. The Chinese have adopted *yan rong* as the nearest equivalent to the European term, karst, since 1966. Before 1966, a Chinese form of the word karst (*Kosute*) was used. Because of the vast size of the country, there is no one phrase or description to indicate karst terrain. As would be expected today, the study of *yan rong* is an important component of Chinese geoscience. Chinese geomorphologists and geologists are proud of their long hydrological records of both river volumes and spring data. These records are held in libraries, like that of the Geographical Institute in Nanjing, and are available for examination by privileged scientists and historians.

All Chinese monographs and papers abound in accurate numerical facts about size, lengths, areas etc. together with attempts to pigeonhole all data into neat categories. Chinese geomorphologists are first-class observers and systematic studies on the dissolution kinetics of calcite or the sedimentary characteristics of limestones, for example, have a high international standard. Many of the modern ideas about karst were introduced into China by scholars who had studied in Europe or N America in the 1920s and 1930s; one of the most influential of such scholars is Ren Mei who studied in Glasgow in the 1930s.

Because many Chinese geologists and geomorphologists are not widely travelled – even in China – there was at first a tendency to explain landscape phenomena in terms of their own local "back-door" experience. Furthermore, because of the language difficulty, only some scientists can read the literature in a language other than Chinese. During the early years of geomorphological study, each group of workers in the different institutes and universities tended to work in isolation and rarely met to compare ideas. However, this stituation has changed rapidly. The isolation is compounded by the immense distances in China; Guilin is nearly 2000 km from Beijing for instance. Each group, though doing good work, tended to pursue its ideas in isolation and to feed on its own mistakes. One of the benefits of the UNESCO/IUGs IGCP project has been that it enabled different groups of karst workers from all over China to come together. China karst geomorphologists are also taking these problems in hand. There are now karst symposia and conferences every year where ideas are exchanged. Setting up a national science foundation on an American model has in the last few years helped to create a much more rigorous and innovative approach to scientific problems.

Throughout Chinese history, earth science knowledge has been used to help land exploitation. This is particularly true in the karstlands. The Chinese were among the first people to adopt irrigation schemes to karst poljes and to develop the use of karst springs. Numerous reservoirs have been built in the karst areas of Guizhou and Guangxi, since the liberation and indicate the skill of the Chinese in karst engineering (Song Linhua 1981; Yuan Daoxian and Cai Guihong 1987). Karst geologists have also given particular attention to the mineral deposits associated with karst-bauxites, and lead-zinc deposits and non-metalliferous resources, such as barytes and fluorite; more recently, the oil–gas reserves occurring in deep karst limestones have received much attention. Thus, Chinese karst study is predominantly applied and pragmatic; this means that the theoretical and academic problems are less developed than in the West, but nonetheless some interesting and illuminating work in these areas has been produced by university groups and state institutions.

Needham, in 1954, attempted to explain why, despite the substantial early achievements of Chinese science, modern scientific methods did not develop in China. As Needham indicated, the reasons for this are mainly cultural and to explain them a depth of understanding of Chinese philosophy is needed that is rarely possessed by the average scientist. Because of centuries of isolation, philosophical thinking became more "closed" compared with modern thought in the West. Particularly since the inception of the "open door" policy in the early 1980s, there has been a rapid development in the exchange of new ideas both within China and with overseas workers.

3 Karst Terminology and Karst Types in China

3.1 Karst Terminology

Before discussing the main karst types in China, some aspects of Chinese karst terminology should be described. In geomorphology, Chinese geographers and hydrogeologists are more concerned with groupings of landforms than with the origins of individual forms. Emphasis in China has been on the hills or positive forms of karst, whereas in Europe more attention has been given to the negative landforms – the closed depressions. Closed depressions interrupt or seem to replace the valley drainage networks, and were regarded by European geomorphologists as the main organizational units in the landscape. The main Chinese terms for different types of karst hills, for dry valleys and for closed depressions are given by Atkinson in Song Linhua (1986a) and to which reference will now be made. These terms are descriptive and do not in any way presume any particular mode or age of formation. There are also some variations in their use in different parts of China.

The most commonly used terms are:

1. The *fenglin* (峰林) is literally a peak forest (a term introduced by Xu Xiake in 1637). In the fenglin, the bases of the hills are all at a similar level and each hill is isolated and separated by valleys, level-floored corridors, or a level plain (Photo 2). The fenglin is often thought of as being most typically developed in the isolated peak area or the tower karst of Guilin; in the Guilin area as in many other areas of tower karst, the fenglin is quite often associated with the allogenic river drainage of the great karst plains and river plains. However, fenglin is also well developed in the plateau areas of Guizhou where isolated conical hills are separated by a network of valleys or closed depressions but where the bases are at a similar level. The separate hills of the fenglin may have many shapes, and many hill angles – rounded cones, pointed cones, isolated steep or vertical towers, for instance; the term is thus not entirely restricted to what in European terminology would be styled tower karst. The fenglin-plain type has also been called fenglin-polje (Fig. 14).
2. The *fengcong* (峰丛) is a peak cluster. In the fengcong there are clusters of several peaks (cones and towers) all sharing a common basement or plinth – not from a level surface. There is often a flat surface between each cluster. Fengcong depression and fengcong valley relief consist of clusters of peaks

with deeper depressions or valleys between each cluster (Fig. 14, Photo 3). Fengcong relief occurs amongst the Guangxi tower karst plains and is widely distributed in the Guizhou plateau – in Guizhou where particularly the *fengcong-depression* and *fengcong valley* types are very common. As stated by Atkinson, "a rough working definition of fengcong is that the depths of the depressions within the clusters should be no more than one-third of the depth of the depressions between them", i.e. a/b = 1/3 (Fig. 14). Maire and Zhang Shouyue refer to *fenglin-ouvala* and *fengcong ouvala* (Barbary 1991).

If uplift of the pre-existing karst platform occurs, then the river valley deepens and forms a canyon produced by rapid downcutting in response to the uplift. New fengcong cones (fluvially formed) are found on the valley sides. This is a *fengcong-canyon* landscape seen in parts of Guizhou and in the Yangtse gorges of Sanxia. The Chinese terminology does not distinguish between tower and cone karst – a differentiation made by Western geomorphologists purely on slope characteristics and irrespective of whether the hills are isolated or clustered. All the normally used Chinese terms and their European equivalents are explained in a new glossary of karst terms in Chinese, recently published by the Karst Institute in Guilin (Yuan Daoxian 1988). A short glossary is also given by Balázs (1989).

The most general term for closed depression is *wadi* (洼地). These are *closed hollows–dolines* or funnels up to a few tens of metres across.

Photo 2. Tower karst (fenglin) *and Quaternary tarraces* along the Lijiang

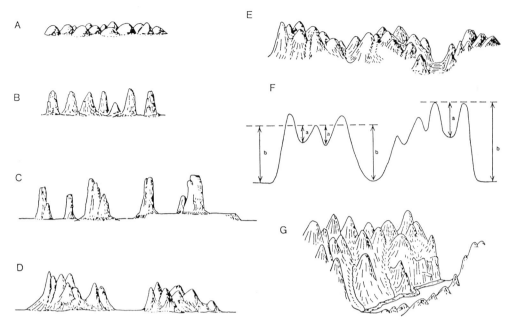

Fig. 14. Sketches to illustrate the meaning of Chinese geomorphological terms (see Sect. 2). **A, B,** and **C** all show different varieties of fenglin – hills whose bases are all at a similar level. **D, E, F** and **G** show fengcong – clusters of hills separated from each other by depressions, valleys or low ground. **A** Fenglin: karst cones, as found in Gunung Sewu, Java and some areas of the Guizhou plateau. The cone or sinusoidally shaped hills are separated by a network of valley and closed depressions. **B** Fenglin: closely spaced karst towers separated by networks of narrow, level-floored corridors. Common in fenglin-basin areas of Guizhou plateau. **C** Fenglin: isolated karst towers rising from level plains. This style of tower karst is usually associated with river drainage of the karst plains. Two levels are shown, the *upper one* representing a river terrace. This style is common in the fenglin parts of the Li river tower karst, Guilin region, Guangxi. **D** Fengcong: clustered peaks sharing a common base, or plinth, with a flat surface between the clusters. Common in the Guangxi tower karst plains and on the Guizhou plateau. **E** Fengcong: clusters of karst cones and towers with deeper depressions or valleys between the clusters. Smaller depressions occur within the clusters. This style is what is meant by fengcong depression and fengcong valley topography. It occurs widely on the Guizhou plateau, frequently in zones around the edges of the undissected parts of the plateau. **F** Fengcong: diagrammatic cross section across a fengcong depression area. A rough working definition of fengcong (as opposed to the fenglin style shown in **A**) is that the depths of the depressions within the clusters should be no more than one-third of the depth of the depressions between them, i.e. $a/b < 1/3$. **G** Fengcong-canyon landscape: produced by uplift of a pre-existing karst plateau, seen on the left skyline of the sketch. The river valley (canyon) has been produced by rapid downcutting, in response to the uplift, developing steep vertical walls. New fengcong cones have formed on the valley's sides. Karst springs may hang hundreds of metres above the valley floor. (Smart et al. 1986)

Photo 3. Cone karst (fengcong), western Guizhou

Closed depressions are hollows larger than dolines, but whose floors are always above saturation level (possibly with short periods of flooding only). They are up to hundreds of metres in diameter. This type of closed depression is the most characteristic *wadi*.

Closed basins are longer hollows, often irregular in plan with flat floors which flood occasionally or regularly. These are more or less equivalent to the European idea of a polje.

Karst valleys are elongated hollows, often several kilometres long, whose floor may carry seasonal streams (within the zone of water-table fluctuations), like the Popovo polje in Croatia. They are seasonally dry valleys.

Karst plains are extensive flat fluvial areas crossed by and drained by perennial streams and are alluviated. The karst plain at Guilin is an example and is covered with unconsolidated deposits up to 20–30 m thick. A karst plain may also have a planed rock surface. Usually, depressions form along the major structural lines and may form between peak clusters, i.e. the fengcong depression. Deepening of the depressions causes them to intersect the water table; they may then be widened by lateral corrosion and erosion into basins.

3.2 Karst Types

Zhang Zhigan (1980a) pointed out that the clear differences in the types of morphogenetic features of Chinese karst are the basis upon which karst

46 Karst Terminology and Karst Types in China

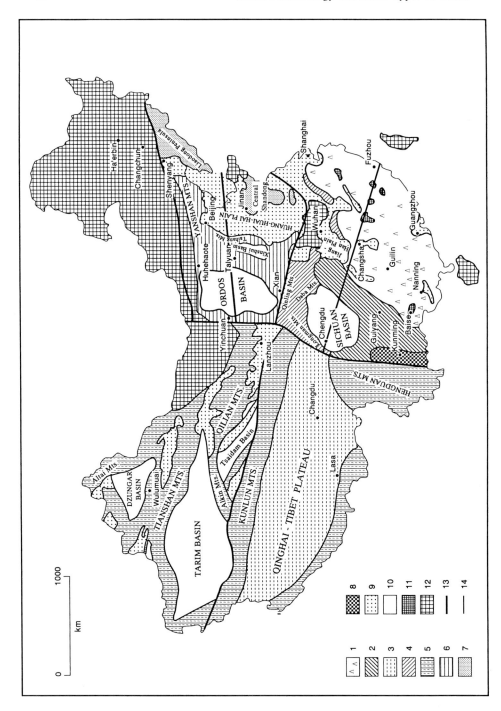

Karst Types

regions may be distinguished. The different associations in time and space between the tectonic units and precipitation zones give rise to various geological-geographical environments and result in the development of the different types of ancient and modern karst landscapes. The moulding of karst landscapes reflects geological processes acting over millions of years and many karst landscapes of China are the products of multistage karstifications. Thus, as indicated in Chapter 1, the main factors which constitute the geological-geomorphical environment in China are *tectonic movements* (rate, amplitude and size and shape of the tectonic block) and *climate*, particularly rainfall and the duration of processes. These are the most important in the moulding of the karst landscapes. These criteria are used in different ways by Chinese geologists and geographers to classify karst types.

The Chinese have made many attempts to bring karst into schemes of climatomorphology and to classify their karst. This has been particulary so since the morphoclimatic work of H. Lehmann on the karst of the Gunnung Sewu, Java (1936a). Many Chinese papers on the classification of karst types in China (notably those by Lu Yaoru, Ren Mei, Zhang Zhigan and Yuan Daoxian 1980, 1982, 1986, 1991) stress the importance of tectonic and climate elements in varying degrees. The classification of Zhang Zhigan is partly used as the basis for karst differentiation in this book (Fig. 15). It is based upon tectonic units, climatic zones and time for morphological development (Table 7). It involves up to 10 main types, with type 11 being the relatively small area of modern coral reef karst in the South China Sea. The framework of the karst types is related to the five major natural landscape regions, described in Chapter 1. The combination of the three major controls of tectonic units, climatic zones and length of time is well illustrated in a section from west to east across China, from the Tibetan plateau to the Guangxi basin (Fig. 6; Zhang 1980a).

In the book, *Karst of China,* Yuan Daoxian and his colleagues (1991) have opted for a classification of karst types on a different basis. These types are based on the simple climatic division of the country into tropical and subtropical karst (south of the Qinling line); karst in the relatively semi-humid and semi-arid N China (north of the Qinling line); and karst in the alpine and high mountainous areas of western China. Each of these major types is further subdivided, quite often on a structural and tectonic basis, so that although the approach is different from that adopted by Zhang Zhigan, the major subdivisions from both authors are much the same. The main problems arise in

Fig. 15. The distribution of karst types in China. *1* Peak forest type; *2* plateau-canyon type; *3* high plateau type; *4* medium mountain type; *5* high mountain canyon type; *6* loess mountain-wide valley type; *7* hill-wide valley type; *8* block lake basin type; *9* buried block mountain type; *10* depressed basin type; *11* degenerate peak forest; *12* areas of igneous and metamorphic rocks; *13* boundary of major regions; and *14* boundary of karst types. (After Zhang Zhigan 1980)

Table 7. Classification of karst morpho-genetic types in China. (Zhang Zhigan 1980a)

Karst-moulding geological-geographical environment			Time for moulding karst	Karst type		Representative area
Tectonic unit		Climatic zone		Name		
Uplift zone	Intense uplift	Massive uplift	High frigid zone	Modern karst	Plateau relict peak-forest type (Qinghai-Xizang type)	Qinghai-Xizang plateau
		Linear uplift	Distinct vertical zoning of high mountain		High mountain-canyon type (Tianshan type)	Tianshan mountains and jingping mountains, western Sichuan
	Moderate uplift	Massive uplift	Humid subtropical and temperate zones		Plateau-canyon type (Guizhou type)	Yunnan-Guizhou plateau and mountainous areas of western Hunan and western Hubei
			Semiarid temperate zone		Loess mountainous area-wide valley type (Shanxi type)	Loess plateau
		Linear uplift	Humid temperate zone		Medium mountain-canyon type (Dabashan type)	Daba, Longmen and Qinling mountains
	Gentle uplift		Humid tropical and subtropical zones	Recent karst	Peak forest type (Guangxi type)	Guangxi basin
			Semihumid temperate zone		Hill-wide valley type (Central Shandong type)	Low mountains and hills in central and southern Shandong, Lioning peninsula and piedmont hills of the Yanshan and Taihang mountains
Subsidence zone	Differential block uplift		Humid temperate and subtropical zone	Ancient karst	Block-lake basin type (Eastern Yunnan type)	Lake-basin area, eastern Yunnan
	Differential block subsidence		Controlled by the paleoclimatic zone		Buried block mountain type (Pohai Gulf type)	Huang-Huai-Hai plain, lower Liache plain, Pohai Sea and Tiang-Han plain
	Depression		do		Depressed basin type (Sichuan type)	Ordos, Xinshui basin and Sichuan basin

central China, i.e. transitional areas from south to north and where the controls of precipitation and tectonics interdigitate, and where the Tertiary and Quaternary palaeokarst are more important in the landscape. Tertiary karst is even more important in N China.

Lu Yaoru (1987a) interpreted the karst types in China in terms of the upwarping rates of the various regions. In an interesting classification he recognizes the following main types:

1. Strong uplift model – seen in Qinghai-Xizang (Tibet).
2. Continuous upwarping model – Yunnan-Guizhou plateau.
3. Differential upwarping model – Sanxia-Huangling Dome (Central Yangtze).
4. Slowly upwarping model – Guangxi and Guangdong.
5. Widespread subsidence model – Tarim, Taihang, Shanxi.

These types, to some extent, can be incorporated into those based on climatic criteria.

3.3 The Main Types of Chinese Karst

3.3.1 Peak Forest (Fenglin) Guangxi Type
(Slowly Upwarping Model of Lu Yaoru)

This is the most famous type and is widespread in southern and southeast China. It is largely a karst of isolated peaks (a tower karst), but does include groups or clusters of hills (fengcong); the isolated peaks rise (usually steep-sided) from a karst plain. The slopes of the individual peaks depend very much upon the structural and lithological characteristics of the limestones. This type is particularly associated with the Palaeozoic limestones deposited upon the Yangtse paraplatform and the S China fold system. It has been caused partly by the allogenic erosion of the great rivers systems of South China, as in the Guilin area. Because of magmatic-induced upwarping, limestones were often not laid down in southeast China or have been removed by erosion, but they have survived in North Guangdong, Guangxi, Hunan, Jiangsu and Zhejiang, almost all south of the Yangtse (Chang Jiang) river. The fenglin karst has developed over a long period of geological time and it is possible that it has been forming since the early Tertiary and even earlier. On this part of the Yangtse platform and the S China fold system, uplift has been slow and has also been accompanied by some subsidence. As a result, the rate of uplift has been equal to or slightly less than the rate of denudation. It is assumed that conditions during the formation of the fenglin have remained basically tropical, but the effects of Quaternary climatic changes have yet to be worked out in South China. However, Shi Yafeng concludes that during the Quaternary, temperatures fell by only 4–5 °C in S China (1992).

There is evidence that during the cold phases of the Quaternary the southern subtropical and tropical parts of China were drier than they are today, i.e. that the influence of the southeast monsoon was considerably less, partly due to the immense high pressures over central Asia and also to the greater expanse of land as a result of the lower sea levels. Uranium dating of stalagmites from the Guilin area shows hiatuses of deposition which correspond to the glaciations in Europe and the Himalayas, indicating that the climate might have been too dry for stalagmitic deposition in the caves (Wang Xunyi 1986). In addition, there are large boulder and mixed superficial deposits in many parts of southern and southeast China – as at Lushan and also the Guilin plain (Yuan Daoxian et al. 1985). Such deposits were originally interpreted by J.S. Lee (Li Siguang) to be of glacial origin; he believed them to be boulder clays formed by glaciers which reached far south into China – Guilin is 24°N. It is now thought that these deposits might be those of debris flows and sludge and mass movements, which flowed down from nearby mountains under highly fluid conditions (Derbyshire 1983). These deposits cover considerable areas and can be quite thick and are of several generations. They may represent periods of great slope movements induced by intense cyclone activity during the Quaternary. There is, therefore, sufficient evidence to show that the Quaternary climates in South China were very variable, particularly in terms of rainfall variations. The effects of these climatic changes upon the development of the fenglin in Guangxi and other parts of South China are as yet unknown, but are probably significantly less than the Quaternary changes in N China. The boulder deposits described by Derbyshire are analogous to the colluvial boulder deposits on granites in Hong Kong and Malay Peninsula (Ruxton and Berry 1957).

The Guangxi type is one of the oldest karst types in China: it is basically a residual karst with residuals sitting either on a non-karst basement (sometimes Devonian sandstones) or on a planed limestone surface as in the Guilin plain. Palaeomagnetic determinations of sediments in caves high up in towers in the Guilin area and well above the present Lijiang river show that the sediments are at least 900 000 years old and may be over 1.6 million years old.

3.3.2 Plateau–Canyon Type (Guizhou Type) (Continuous Upwarping Model)

This area is still part of tropical and subtropical China, but is separated from Guangxi and the lower areas on a morphotectonic basis. This type is located in the Yunnan–Guizhou plateau and in the mountains bordering the provinces of Hunan, Hubei, and Sichuan. It is characterized by two highly contrasting landscape units – the canyon area and the plateau area. This type has been formed by a relative rapid rate of uplift – possibly from the middle or late Pliocene through to the Quaternary in response to the neotectonic movements associated with the Himalayan movements and the uplift of the Tibet

(Xizang)–Qinghai plateau. In the late Neogene and the Quaternary, the rate of uplift exceeded the rate of karst denudation, so that the area was subjected to intense downcutting and headward erosion and corrosion by the surface and subsurface river systems. It is believed that, in the early Pliocene, the Guizhou-Yunnan plateau was a relatively flat and eroded mostly limestone surface which developed under warm conditions. The major tributaries of the Changjiang (Yangtse) and Zhujiang (Pearl) rivers now rise on the Yunnan–Guizhou plateau and cut through the plateau creating deeply dissected canyons up to 700–1000 m deep. The altitude of the Guizhou-Yunnan plateau is 1000–2000 m a.s.l. It is believed that most of this uplift was late Tertiary, but that there was also Quaternary uplift of at least 300 m. The Sancha he river for example is cut over 500 m deep into the Guizhou plateau. The uplift has inaugurated a new phase of karstification which is now destroying the preexisting planation surface, and the bases of the canyons are now the base levels. Where the new phase of karstification has not reached the plateau, the pre-existing planation surface is preserved along with the ancient hydrographic networks still acting as base levels. A zonation of landforms from watershed to river valley is preserved.

The relief on the plateau surface is relatively gentle. Where the limestones are exposed at the surface, as in much of Guizhou, the plateau is characterized by many different types of cone hills with gentle and steep sides – the shapes of the cones depend particularly upon the lithological characters of the limestone. The cones can be isolated, forming a cone-type fenglin or in clusters forming cone-type fengcong. In other parts of the plateau surface, especially in Yunnan, the limestones are covered by more recent sediments, often Eocene clays – a few tens of metres thick. Where these clays are being stripped off as at Lunan in the stone forest (Shiling), modern karst processes are at work changing karst features which originally formed beneath the clay cover. In parts of East Yunnan, the plateau surface has been affected by recent tectonic movements; block-faulting has formed elevated eroded blocks (over 2000 m a.s.l.) and down-faulted lake basins containing Quaternary sediments, as at Dianchi lake, near Kunming.

The Guizhou karst type lies between the strongly uplifted Tibet plateau (over 3000–5000 m) and the lower Guangxi peak forest karst plain. The boundary between the Guizhou plateau and the Guangxi plain is marked by a break of slope (Fig. 6). A mixture of the Guizhou–plateau–canyon landscape and the Guangxi peak forest occurs on this slope, which is well seen at Dushan (Song Linhua 1986a).

3.3.3 High and Medium Mountain Canyon Type (Differential Upwarping Model)

China has many mountain ranges in which karst rocks are important. In almost all of these areas, rapid vertical uplift producing strong relief has occurred in the Tertiary and probably Quaternary. Distribution of karst in the high mountain areas is fractured and discrete, because of the dissection by gullies and valleys along structural lines. Limestones occur in most of the high and medium mountain areas, including the Chinese Himalaya (S Tibet), the Hengduan mountains (Yunnan), the Qinling mountains, the Longmen mountains (in Sichuan) and in the Kunlun mountains and other mountains in Chinese central Asia. As a result of the variations in latitude, rainfall and height, these areas have a varied group of karst landscapes.

In many of these areas of high altitude, karstification is sporadic, because of the importance of frost action at high altitudes and gully erosion. Frost and other mechanical weathering with strong mass movement due to gravity in the steep relief all combine to outweigh the effects of solution and karstification. The weathering of the limestones is, therefore, similar to that of the other rocks and because of the susceptibility of limestones to frost weathering they often form relief lower than other rocks and are accompanied by large talus deposits of limestone fragments. Surface karst phenomena are relatively rare except where local snow and ice conditions and water conditions give rise to solutional fissures and karren. Rain pits, grooves and snow patch solution occur where meltwater is important. Meltwater also gives rise to caves and vertical subkarst features can be well developed. Karst water may circulate at depth in the deeply dissected valleys often to many hundreds of metres. The density of karst conduits is very low, however, and deep karst very unevenly developed. Because of the dissection of the mountain groups by valley and gullies along faults and thrusts, the exposures of the limestones are fractured and cut up into relatively small outcrops; thus, any limestone catchment areas are very small. Areas of glaciated limestone pavement are quite restricted; this is because of the relatively small areas of glaciation even in N and W China. This is the considered opinion of most scientists who have worked in West China. However, Kuhle (1987) has argued for an extensive ice cap on the Tibet plateau.

This type of karst exists in many climatic and precipitation zones (Fig. 15). The wetter areas, such as in the Himalayas and in the Hengduan mountains (West Sichuan), mostly resemble an alpine-type karst or a Rocky mountain Crows Nest area karst (Ford 1979). Conditions in China are, however, drier than in the Alps and the Rockies and because of the lower latitudes, they are warmer in summer. In the drier areas of the Kunlun and mountains north of the Qaidam basin, conditions are very arid (less than 25 mm annual precipitation in places). The karst in these areas is basically a high altitude desert karst, similar to that occurring in Saudi Arabia or parts of Namibia (Felber et al. 1978).

Other areas of high and medium mountain type are wetter and more karstic, with well-developed solutional features. The karst of the San Xia (Three Gorges) of the Yangtse (Chang jiang) belongs to the medium mountain karst type. In its passage from the Sichuan basin into the East China plain, the river cuts through both the Ordovician platform zone and the later

Mesozoic red beds; the karst is transitional from a medium mountain type to the plateau-canyon type. As suggested by Lu Yaoru (1987a), the western parts of the gorges (Qutang xia and Wuxia) have been strongly uplifted and are high mountain karst, but the easternmost gorge (the Xiling xia) has suffered less uplift and has more the characteristics of a plateau karst.

Medium high mountain type is also found in E Sichuan where NE-SW trending folds form parallel mountain ridges and valleys. Lower Triassic and lower Permian limestones occur in the axes of the steep, narrow anticlines and form long karst valleys up to 110 km long. The karst is undulating with big closed depressions, sink holes, underground streams and many caves. In the lower Permian Maokou limestones, there are several "gigantic karst collapse pits" (Yuan Daoxian 1991). West of Hangzhou (Zhe jiang province) in E China, some of the relief might be regarded as of the peak cluster type; it is developed in Lower Palaeozoic platform limestones. There are also substantial limestone formations in the southern Qinling where landforms similar to peak cluster occur.

3.3.4 Periglacial Type (Tibet Plateau Type) (Strong Uplift Model)

The Tibetan plateau is the largest area of high plateau on the earth's surface, having an altitude of 4500 to over 5000 m a.s.l. Bands of Upper Palaeozoic and Mesozoic limestones outcrop in wide and generally E-W trending outcrops. The rock outcrops are covered with frost-shattered limestone fragments, with periglacial pinnacles piercing the debris cover. The present annual average rainfall is about 300–400 mm in the Lhasa area, and therefore, modern karst features are restricted to rudimentary karren and occasional small caves (usually less than 10 m long). The permafrost and the dry climate generally prevent the development of subterranean drainage, long river caves and springs. However, deep-seated thermal springs associated with the recent and present neotectonic activity in Tibet occur in the deeply faulted areas of the plateau. The relief on the Tibet plateau is of the order of 200–400 m and is relatively rolling country, but is reduced by the frost weathering to gently sloping debris fans, interrupted by frost-shattered rock pinnacles. Rainfall in W and N Tibet is less than 100 mm/year.

3.3.5 Hill and Wide-Valley Type (Central Shandong)

This type is what would often be called a "temperate" or subhumid (semi-arid) karst – familiar to workers in Europe and North America. However, the writer has deliberately refrained from using the terms temperate and Mediterranean to avoid climatic connotations, particularly temperature correlations. However, climate may be important in the formation of these types, in that, the precipitation in N China is so much less and with fewer heavy storms which give rise to the rapid runoff considered important in the development of the peak and cone karst of S China. Central Shandong has about 500–700 mm/year precipitation. Karst of this type occurs not only in West and East Shandong, but in the isolated limestone masses southwest of Beijing, like Choukoudian (Peking Man's cave) in Ordovician limestones and in the Liaoning peninsula northeast of Beijing. These areas covered by so-called normal karst are not nearly so extensive as the peak forest and cone karst of South China.

The central Shandong area is formed of platform limestones (Ordovician and Cambrian) which dip off the Sino-Korean platform, having been removed from E Shandong. In Liaoning and Nei Mongol, Ordovician and Proterozoic limestones are involved. The relief is similar to hilly and low mountainous limestone areas in central Europe with dry valleys and palaeokarstic features being very important. Small, closed depressions and superficial solution features occur. Caves and underground streams occur in central Shandong. Because the Ordovician limestones are highly fissured, i.e. quite karstified, large springs are important as at the Baotu spring in Jinan. The Majiagou formation in the Middle Ordovician in Shanxi is particularly well karstified, and also the Wumishan dolomite of the late Precambrian. The presence of sulphates (gypsums) interbedded in the Majiagou carbonates is very important in Shanxi.

The Zhoukoudian area contains the caves which have yielded Peking Man; it is situated in hilly Ordovician limestones which are much fractured and folded. The caves are fragments of earlier, much larger caves and are much denuded. The same can be said of the hilly remnants of the Ordovician limestones.

In the Liaoning and Jilin areas north of Beijing, the limestones were deposited in the Sinian-Yao sedimentary basin and are Precambrian and quite dolomitic. These dolomites are overlain by Cambrian and Ordovician limestones. Some quite big caves and underground streams occur in this area, near Dalian and Shenyang. Rainfall is over 800–1000 mm.

3.3.6 Loess Mountain Valley Type (Shanxi Type)

Rocks of Sinian, Cambrian and Ordovician age underlie large parts of Shanxi and W Hebei. Situated on the North China platform, they are 1000 m thick and cover an area of 50 000 km^2. The rainfall is relatively low – only 400–500 mm and less, decreasing to the NW, so the climate is verging on the semi-arid. What makes the Shanxi area so different is that the limestones are covered by Quaternary loess deposits 10 to 100 m thick. Dry valleys and large karst springs are the predominant karst features with little development of surface solutional landforms. The underground fissures are enlarged by slow seepage solution of the surface water, resulting in a dense network of solution fissures in the rocks. The surface landscape is composed of mountains and wide gentle valleys; the mountains are loess covered, while the valleys are filled with reworked loess and alluvial deposits. The surface and subsurface karst which developed in the Tertiary is now preserved under the loess and other superficial deposits; though as the loess is removed, the palaeokarst features formed earlier, are rejuvenated.

The landscape is, therefore, dominated by the loess – with a complete absence of karren, and few closed depressions, and very short caves (under 10 m long). On the summits of the mountains where residual Tertiary planation surfaces survive, the old fossil or palaeokarstic depressions and cave systems are preserved. However, normally, mechanical denudation of the limestone is over 30 times that of carbonate corrosion. In sections of weathered limestones at Dajingou, Shanxi, the sections are characterized by (1) a dominance of mechanical weathering over chemical weathering – there is a cover of limestone detritus; (2) a calcareous crust (calcrete) just below the soil layer, and due to the semi arid climate, carbonate deposition replaces solution.

Rainfall is concentrated in this area from July to September. The intense mechanical weathering provides large quantities of detrital material; this accumulates in the flood channels and so the wide riverbeds are filled with up to 200 m of gravelly material (poorly sorted and rounded) and the longitudinal gradients of the valleys are very gentle. Both this debris and the loess have buried the old Tertiary karst; buried depressions have been found by drilling in the river valleys – as in the Songxi valley, in central Shanxi. From northwest to southeast across the area, the climate becomes more humid and the loess deposition weakens, and the intensity of karst development increases, so that SE of Shanxi, in the Majiagou limestones, the karst is beginning to be more like that of West and central Shandong.

In general, the present degree of karstification of the soluble rocks in N China is lower than in S China. However, a few individual strata are intensely karstified – notably the Majiagou limestone of the Middle Ordovician, which is also highly permeable. The Majiagou is a mixed sulphate and carbonate formation – made up of alternating gypsum-bearing rocks, 20–50 m thick and carbonate limestones 100–200 m thick. The combination of the easy dissolution of the sulphate rocks and the good water permeability of the carbonates means that the karst development in the Majiagou is intense. It gives rise to large karst water basins whose outlets are large springs. Large caves and undergound rivers, other than in the palaeokarst, are seldomly encountered – very different from the Guangxi and Guizhou karst types. Instead, the underground karst features are dominated by corroded fissures which are usually closely spaced to form a zone of corroded fissures.

3.3.7 Buried Karst or Deep Karst (Widespread Subsidence Model)

Ren Mei et al. (1982) referred to this type of karst as deep karst. The problem of karstification and permeability of deep carbonate strata is an important aspect of karst study. This is because deep karst rocks provide reservoirs for oil, gas and geothermal hot water. The buried karst in China is of two types– the buried block mountain type (Bohai gulf type) and the depressed basin type (Sichuan type); in the latter, the soluble strata have not been exposed at the surface since the Mesozoic.

3.3.7.1 The Bohai Gulf Type

The area of the Bohai sea consists of buried block-faulted structures which have been buried by Cenozoic deposits. They may be kilometres in height and up to 2000 km^2 in area. They lie in the subsidence zone between the Taihang, Yanshan and Liaodong peninsula and central to South Shandong. The karst was formed in Proterozoic and Cambro-Ordovician limestones in subsidence areas. The block faulting is not unlike the differential block activity, at present active in East Yunnan, however, these basins in Yunnan are now exposed.

In Bohai, the stage of block mountain formation is Eocene and Oligocene. By the end of the Oligocene a series of an echelon block mountains (horsts) and basins (grabens) had been produced. The block mountains subsided and were buried at the beginning of the Miocene. By the Holocene the top of the mountains had been buried to a depth of 1–4 km. The early Oligocene basin deposits contain subtropical plants, suggesting a periodic subtropical climate for the Bohai. It is believed that this was the main epoch of the karst development in these buried block mountains.

The oil field development reflects the main high-yielding formations of the Wumishan dolomite (Proterozoic), the Gaoyuzhuang dolomite, Majiagou limestone and the Fujunshan dolomite. In the buried mountains, the karstification of the dolomites appears to be more intense than that in the limestones, which is the reverse of that nearer the surface. The buried mountains of the Bohai gulf have experienced three main stages of karstification, and the duration of the karstification has a marked effect upon the permeability of the rock masses. The three main phases occurred in the Caledonian, Yanshanian, and the Himalayan stages; the strata which have undergone all three stages of karstification have the most intensive karstification and the greatest permeability, so they are the main reservoir formations in the area.

3.3.7.2 Sichuan Buried Karst Type

The Proterozoic-Triassic limestones which underlie the Sichuan basin were covered by up to 1000 m of red Mesozoic sandstones (Jurassic and Cretaceous), and have never been exposed. The primary state of deposition is perfectly preserved. The main karst features were moulded during the stage of deposition – and form a palaeo- or ancient karst, evolved in a generally closed chemical environment. The buried limestones of Sichuan form important reservoirs for oil and gas. In Pengshui, SE Sichuan, there are salt and hot springs and fresh water springs; there is much mixing of the different groundwaters. A cave 4.45 m high was drilled through at a depth of 2900 m (2400 m below sea level) in the Sichuan basin. Buried karst in Ordovician limestones also occurs in the Tarim (Talimu) basin at a depth of 5000 m.

The distribution of the main karst types in China (Fig. 15) illustrates the combination of the effects of climate, with the uplift and subsidence of the earth's crust in recent geological time. Uplift and subsidence not only dominate the changes in base level of water discharge, and hence determine the characteristics of the hydrographic networks, but also affect the combination of the climatic factors and the processes of exogenous agents through the changes in the relief (Zhang 1980a). It is also of interest to note that, compared particularly with western Europe, glacially modified karst is very restricted in its occurrence in China. This is the result of the generally much more restricted actual glacial occurrences in this part of Asia in the Pleistocene, even though the cold phases of the Quaternary were extremely cold. Periglacial karst, on the other hand, is very well developed.

4 The Guilin Karst

4.1 Introduction

The name "Guilin" means "the forest of the sweet-scented osmanthus tree" (*Osmunda* sp., *Osmanthus delaboyi*) and the town is one of the historical cities designated by the Chinese government. The town goes back to the Qin Dynasty (221–206 B.C.) and was named Shian under the Han Dynasty (111 B.C.). As the political centre for the Guangxi region in the Tang Dynasty (618–907 A.D.), it has many stone inscriptions and statues of that age, and continued to be predominant under the Song Dynasty (960–1279 A.D.), becoming the provincial capital under the Ming and Qing Dynasties. During the Ming Dynasty in 1393, Prince Jing Jiang constructed a palace with pavilions, terraces, altars and temples. Although damaged in the Qing period, the palace was sufficiently repaired for Sun Yatsen to set up his headquarters there in 1921. It now houses the Guangxi Teachers' University (Ru Jinwen et al. 1991). During the Sino–Japanese war, when much of China was occupied by the Japanese, many scientists, artists and writers fled to Guilin. Among these was J.S. Lee (Li Siguang), the famous geologist whose laboratory was at Yanshan, S of Guilin.

Mention of Chinese karst in any geomorphological discussion of karst leads naturally to thoughts of the karst of the Guilin area. This area is also regarded as one of the most beautiful and spectacular in China and the boat journey down the Lijiang from Guilin to Yanshuo is one of the most scenic in the world (Fig. 16, Photo 4), Guilin is situated in the NE part of the Guangxi Zhuang Autonomous region at about 25°N latitude. There is a great variety of landforms in the Guilin area, but it contains also some of the oldest karst features in China and is one of the most complex landscapes to unravel. It is situated at about 140 m a.s.l. on the third great step or plain level of the Guangxi basin in the Chinese landscape (Zhao 1986). It also lies towards the eastern extremity of the S China karst areas; to the east towards Guangzhou more of the limestones have been removed by denudation or were never deposited and the karst, though still of interest, is becoming less continuous and more isolated in nature. The average annual rainfall of Guilin is nearly 1870 mm and strongly seasonal, the mean annual temperature is 19.7 °C. The southeast monsoon rainy season in South China begins in early April and there are further rains later in July and August. As in other parts of China, the seasonal distribution of precipitation is not homogeneous (Zhao

Introduction

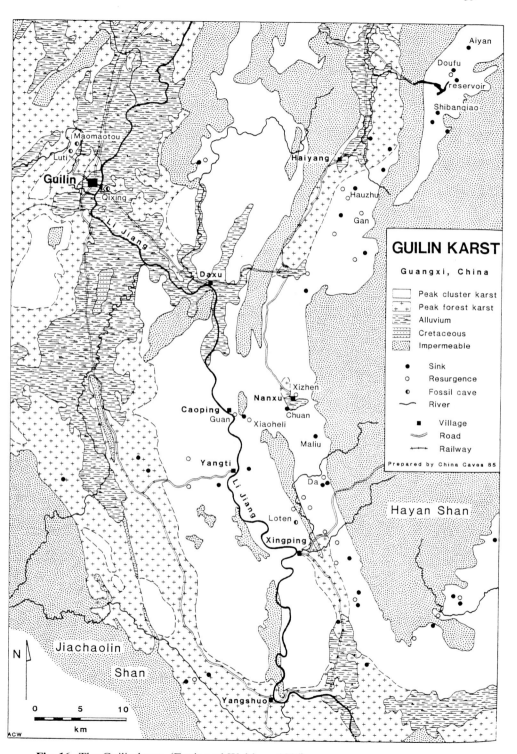

Fig. 16. The Guilin karst. (Eavis and Waltham 1986)

Photo 4. Tower karst (fenglin) and the Lijiang gorge, near Yangti

1986). Though karst similar to the Guilin fenglin occurs in other areas of South China, the wide range of literature available to Western readers on the Guilin area makes it appropriate to treat the area separately in this book. Despite the problems concerning the age and the complexities of the Guilin karst, it seems the most obvious area to begin any discussion of the landforms.

The area usually regarded as the Guilin district is about 130 km from north to south and from 20–50 km from east to west. The karst is developed in Middle and Upper Devonian limestones and part of the Lower Carboniferous. The Guilin basin, based on the S China Caledonian fold system, is a composite synclinal basin, trending in an arc-like form generally from N-S for about 110 km. It is frequently referred to as an epsilon structure. The convex part of the arc faces westwards (Fig. 17). The beds are fairly steeply dipping on the limbs of the syncline but in the axis are quite gently dipping. As shown in Fig. 17, there are many generally trending N-S strike faults, but the rocks have also been cut by later transverse (E–W) faults and shear faults (NW–SE). The basement on which the Devonian rocks were laid down is the Cambrian-Silurian – these rocks were folded and consolidated before the limestones

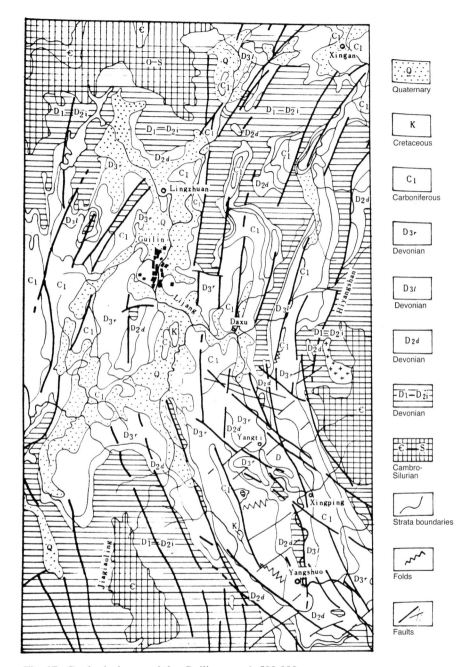

Fig. 17. Geological map of the Guilin area, 1:500 000

were deposited – and also formed the basement controls when the limestones were later folded. It is believed that the arc-like structure of the Guilin area was formed partly during the Variscan, but also mainly during the Indo-Sinian and Yanshanian movements of the Mesozoic, possibly under the influence of the motion of the Pacific plate. Although the Tertiary Himalayan movements did not affect Guilin to the extent that they affected Guizhou to the west, there is substantial evidence that considerable early Tertiary upwarping occurred and that this upwarping was important in the evolution of the karst landforms. In the later Tertiary and Quaternary, tectonic uplift was slower, though continuous.

Over 2600 m of Middle and Upper Devonian and Lower Carboniferous limestones are exposed in the Guilin syncline. They consist of a platform facies and a basin (trench) facies and they cover 76% of the total area. The main karst-forming rocks are of platform facies and are shown in Table 8 (Weng Jintao 1987a).

The Lower Devonian in the Guilin area is composed of clastic rocks – violet-red/green sandstones, and shales – and the lower Middle Devonian is formed also in clastic-sandstones and shales. These outcrop in the mountainous areas on either limb of the syncline and in minor anticlines within the major composite syncline, and are important in the development of the karst. The main limestone beds are presented in Table 9.

Outcrops of later red beds lie unconformably upon the limestones in parts of the Guilin area – the largest exposure being near Guilin airport. The base of these red beds is conglomeratic and brecciated containing fragments of limestones which indicate that, at the time of its formation, mechanical erosion of the rocks was predominant. The basal breccias are found at many altitudes

Table 8. The karst-forming facies in the Guilin area. (Weng Jintao 1987a)

		Platform facies		Trench
		European	Chinese	Chinese names
Carboniferous	Lower	Visean	Huangjin Formation	
		Tournasian	Upper Lower	
Devonian	Upper	Famennian	Rongxian	Luoming Formation Wuzhishan Formation
		Frasnian	Guilin Formation	Lazhutai Formation Fuhe formation
Devonian	Middle	Givetian	Donggangling	Donggangling Formation

Table 9. The main limestone beds in the Guilin area. (Weng Jintao 1987)

			Thickness (m)
Upper middle Devonian	Dongangling formation	Dark grey biomicritic limestones	30–485
Upper Devonian	Guilin formation	Dark grey biomicritic limestones	410–1606
	Rongxian formation	Pale grey oosparitic	387
Lower Carboniferous	Lower, Datang Upper formation	Grey micritic limestones (or dolomitic limestones or dolomites)	620

on the limestones. Rb/Sr dating of the illite in the red bed series indicates that they belong partly to the Triassic and partly to the Cretaceous; this is supported also by faunal evidence (Weng Jintao 1987b).

Along the Lijiang and in topographic basins, there is a series of Quaternary deposits – mainly yellowish-red clays, with sandstone boulders of all sizes – derived largely from the flanking Devonian sandstones. All carbonate boulders are quickly removed by solution. The origin of the clayey-boulder deposits is still disputed, but they are most likely to be debris and mud-flow deposits, as described by Derbyshire in Lushan (1983) and by Bull and others (1989) in Guilin, probably in wet phases of the Quaternary. Sediments occur in the caves and have been dated by various means and discussed by Bull et al. (1989) and Williams et al. (1986). Holocene gravels and sands have been laid down in the river terraces of the Lijiang. The main terrace of the Lijiang is between 3 and 8 m above mean river level – though some workers suggest that isolated hills at 20–40 m are also relics of a higher terrace. An interpretation of the superficial deposits in glacial terms is given by Wang Kejun et al. (1988).

From a geomorphological point of view, Guilin city is most fortunately situated. Above Guilin, the valley of the Lijiang is wide and is flanked mainly by the sandstone hills, with only isolated limestone residuals (Photo 5). However, at Guilin, the limestone hills become more numerous and the Lijiang below the city to the SE enters the famous gorge at Daxu where it cuts through the karst of the Devonian limestones for over 60 km and is over 300 m deep. The gorge trends generally NNW–SSE. The reasons for the origin of the Lijiang gorge are still not known; a wide valley of Quaternary deposits exists south of Guilin city which would seem to be the original valley. There are two main possible explanations for the diversion: (1) that the great thickness of

Photo 5. Guilan city, fenglin plain

clay and gravel deposits which must have characterized certain periods of the Quaternary were of sufficient thickness to divert the river south eastwards into its present gorge (Lin Jinrong 1992); (2) tectonic movements affected the Western limb of the syncline and caused the river to migrate up dip; later, the eastern limb of the syncline was differentially uplifted. The river appears to be antecedent and the Lijiang in the gorge has some conspicuous and large meanders; it must have been incised and fixed from a more gentle relief than it cuts through now. There are well-marked Holocene and late Quaternary terraces in the gorge. There is no estimate of how long these would take to form; but if the diversion was early Quaternary, then there would seem to be sufficient time for the present terraces in the gorge to develop. Quaternary deposits at Nanxu, E of the Lijiang, have been dated by ^{14}C from carbonized wood as old as 37 000 ± 500 years (Ru Jinwen 1981; Ru Jinwen et al. 1991). Furthermore, the underground stream connections, particularly from the sandstone hills to the E, are quite well developed, for example, the Guayuan cave between Nanxu and Caoping (Fig. 16). Whatever explanation is correct for the present course of the Lijiang, the line of the present river, which closely follows structural trends, must have been a well-defined physical line of weakness before any diversion or before it became fixed in its present gorge.

The karst of Guilin is made up of peak forest (fenglin) and peak cluster (fengcong). Much of the Guilin area consists of isolated, steep limestone residuals rising from the planed limestone surfaces, covered quite often by recent alluvium or Quaternary deposits, i.e. it is classical tower karst. Waters

Introduction 65

Photo 6. Lijiang plain, with fengcong on the west side, near Yangti

draining from the Lijiang and the smaller rivers from the sandstone hills flow into the tower hills by foot caves and steepen the residuals. During the rainy season, the rivers bring down pebbles from the sandstone areas. The climatic variations in rainfall – from dry to wet season – are important; over 60% of the annual precipitation of 1870 mm falls from April to August.

In very general terms, *fenglin* occurs in the centre part of the syncline, particularly where the dip of the rocks is low (Figs. 16, 17; Yuan Daoxian 1985a). The *peak cluster* or *fengcong*, the clusters or groups of generally conical hills rising from a common basement tend to occur on the flanks of the syncline or at the periphery of the basin. Moreover, the Lijiang in the gorge section between Daxu and Yangshuo flows almost entirely across a peak cluster even though much of this is at a high level (Photo 6). This fact again indicates that the present course of the Lijiang is a relatively recent one and that the influence of the allogenic water has not yet had the time to change the nature of the peak cluster. Western observers would identify peak cluster as close set cone karst. In the Guilin area, as in many others, peak forest and peak cluster interdigitate, but have quite different hydrological characteristics and with different cave patterns. It is believed that the peak forest develops only where the water table of the phreatic zone is near the surface and where external allogenic water floods the karst, while the peak cluster occurs where the vadose zone is very thick and the water table deep – particularly in zones of uplift. It is therefore found on mountain slopes and in recently rejuvenated karst zones. It was at one time thought that a peak cluster would evolve over time to peak forest and that the two types of landforms formed an evolutionary sequence. There is much evidence to suggest that this is not always so – particularly because of the complexities of their distribution (Yuan Daoxian 1985a) and the relationship of the peak forest to the flooded basins. It is possible to suggest that they have separate evolutionary histories. However, if for any reason the peak cluster comes within the zone of the water table, then its character will become more like the peak forest. A series of block diagrams showing Zhu Xuewen and Zhu Dehao ideas on the development of peak forest and peak cluster in the Guilin area is given in their book (1988) and also in *Guilin Karst* by Zhu Xuewen (1988a; Fig. 18).

4.2 Peak Forest (Fenglin)

A detailed consideration of the two main types of landform in the Guilin area will show the problems associated with them. The steepness and form of the isolated towers or peaks depend upon the nature of the limestones and their lithology. Thus, from a detailed study of the carbonate rocks of Guilin, Weng Jintao recognizes five major genetic types of carbonate rocks (Weng Jintao 1987a). These genetic types are the sparry allochemical limestones, micrite and micritic allochemical limestones, leopard lime–dolomite, organic micritic dolomite limestones, lenticular pelitic limestones; medium crystal dolomites. According to Weng, each one of these major types gives rise to a major peak form (Table 10). He claims that the very steep isolated peaks occur only in the sparry allochemical limestones which characterize the Rongxian limestones in particular: the mean porosity of the Rongxian limestones is 0.68%. He further devises two indices – *isolation degree* defined as the percentage of peak area or unit area of the rock, the smaller the percentage the more the

Table 10. The textures of the carbonate rocks in the Guilin area. (Weng Jintao 1987)

Genetic texture of carbonate rocks	Peak forms
Sparry allochemical limestone	Tower and cluster, saddleback
Micrite and micritic allochemical limestone, intercalated with dolomite	Monocline and taper, spiral
Leopard lime-dolomite, organic micrite dolomitic limestone	Taper and spiral
Lenticular pelitic limestone micritic sparry allochemical limestone	Spiral and taper
Medium crystal dolomite	Dome

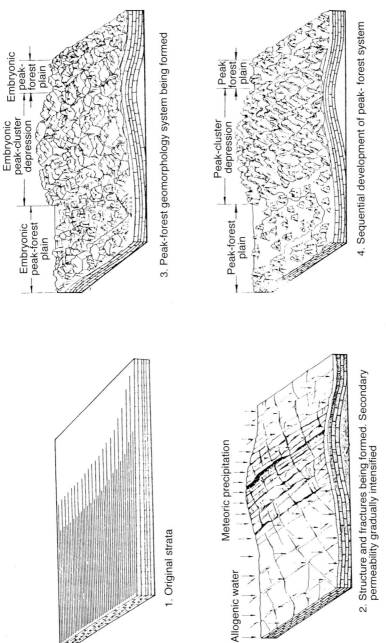

Fig. 18. Development of fengcong and fenglin according to Zhu Xuewen (1990)

isolation degree; and the *cavity degree* defined as the percentage of cavern area in a unit area of peaks, the more the percentage, the more the cavity degree. Calculations on 263 isolated peaks in the Guilin area suggest that isolated peaks consist predominantly of sparry allochemical limestones. Further, under the same allogenic water conditions, mountains composed of sparry limestones are four times more cavernous than the mountains composed of micrites – despite the micrites having a greater corrosion rate and covering an area nearly three times as great as the sparites (Tables 11, 12). Cavity degree governs the slope of the peaks – the greater the cavity degree, the smaller the area and the steeper the slope. Isolated peaks rarely occur in micritic terrain (Photo 7). All these calculations are based on the composition, textures, corrosion rates, physico-mechanical properties, stratigraphical occurrence and thickness of the beds, joints and fissure intensities in the rocks (Weng Jintao 1987a).

Zhu Dehao in Yuan Daoxian et al. (1991) made a morphometric survey of 220 fenglin towers in the Guilin area which revealed the following:

Average height of karst tower: 74 m
Ratio major/minor axis: 0.197–1.0 average 0.66
Diameter of bottom plane: 59–700 m, average 208 m
Direction of major axes: 70°–90° and 290°–300°
Density of towers: 1.23 to 1.59 per km^2.

Table 11. Isolated peaks in the Guilin area. (Weng Jintao 1987a)

Rock type	No. of isolated peaks	Percentage number
Dolomite	9	3.41
Micritic allochemical limestone	32	12.12
Sparry allochemical limestone	172	65.15
Microsparitic limestone	50	18.94
Total no. of peaks	263	

Table 12. Peaks and caves in the Guilin area. (Weng Jintao 1987a)

Rock type	No. of peaks	No. of caves	Cave length (L) (m)	Cave volume (S) (m^3)	Cavity degree $k_2 = S/m \times 100\%$
Micritic allochemical limestone	21	71	3 745	27 901	0.90
Sparry allochemical limestones	43	182	13 460	92 803	3.97
Microsparitic limestones	10	25	2 617	40 363	2.99

Peak Forest (Fenglin)

If the indices of Balázs are applied to the karst towers of the Guilin area, then 60% of the Guilin towers belong to the Yangshuo and Oreganos types in the Balázs classification (Balázs 1973).

The biggest caves, such as the Reed Flute cave are developed mainly in the sparry allochemical limestones. Maze-like caves occur in the dolomitic limestones. The development of big caves and vertical (near-vertical) isolated peaks in the sparry limestones occur in a rock which has a lower corrosion rate than the micrites or biochemical limestones. The resistance of the sparry limestone is partly due to the greater thickness of the individual layers, and the far-extending vertical jointing (directing corrosion vertically): it is also due to their greater physico-mechanical strength, i.e. low permeability and porosity (0.67%) and greater tensile, compressive and shearing strengths (Table 13). Where the sparry limestones occur in the axis of the Guilin syncline and the dips are gentle, the ideal conditions for steep isolated peaks and big caves occur. The big caves and steep peaks depend not only on the corrosion rate but on the strength of the rock to form a framework for the caves and peaks. The well-developed jointing favours the discharge of the water in the initial conduits which progressively enlarge, and this is accompanied by mechanical breakdown to form gallery or hall-like caves. Collapse is also an important mechanism at the foot of the steep peaks – and can be seen often – especially where undercut by streams; solution of the collapsed blocks takes place and the steep peaks retreat continuously. Big collapse blocks occur at the base of many peaks (Figs. 19a, b).

Photo 7. 'Taper' peaks and the Lijiang at Yangshuo

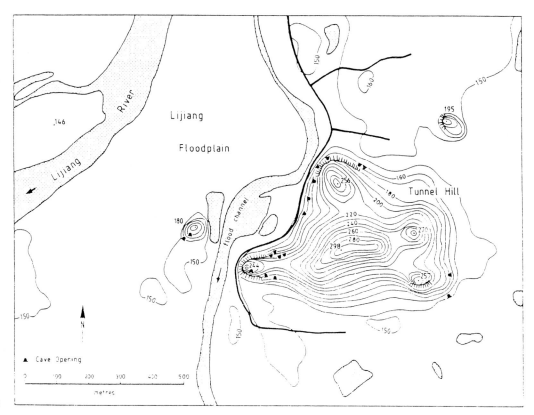

Fig. 19. a Tunnel hill, a karst tower near Guilin, China; *black line* roads and trackways. (Williams et al. 1986) **b** Caves in the Yin hill cave system; *laddered areas* artificial stairways. (Cave Research Group of the Institute of Karst Geology 1983)

Table 13. The strength of limestones in the Guilin area. (Weng Jintao 1987a)

Rock type	No. of sample groups	Average compressive strength (kg/cm^2)	Average tensile strength (kg/cm^2)
Dolomite	9	1139	32.7
Micritic limestone	9	1331	34.0
Sparry allochemical limestone	10	1127	36.8

New explanations of the great variety of peak forms in the Guilin area have been made since 1980. Many of the peaks in Guilin city are in the sparry allochemical limestones of the Rongxian formation – for example, Fubo hill, Elephant Hill. South of Guilin in the axis of the syncline, vertical towers in sparry limestones occur near Putao. The micritic limestones being more thinly

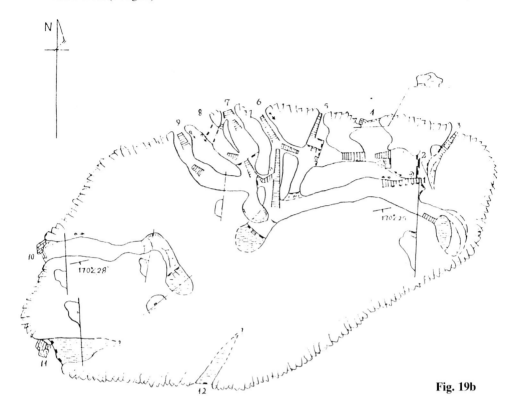

Fig. 19b

bedded give rise to asymmetrical peaks with gentler dip slopes and steeper scarp slopes and are well seen in the Guilin formation near Guilin. The famous Lotus Peak Biylian at Yangshuo is defined by Weng as a taper peak (or cone) (Photo 7). It is developed in leopard porphyritic lime-dolomite from the Donggangling formation. Spiral peaks are developed in bedded micrite-allochemical limestones also in the Donggangling formation. Dolomites are common in the Yanguan stage of the Lower Carboniferous. Although dolomite has a low corrosion rate, it has a high porosity and when slightly weathered shows a rapid increase in porosity (up to 11%) and a decrease in strength. It is, therefore, subject to both corrosional dissociation and mechanical disintegration giving rise to collapse of the individual particles of the rock. Weng (1987a) calls this "en masse" karstification of dolomite and the dolomites form low gentle conical hills, covered with thick dolomite powder (Photo 8); as seen near Ertang, SW of Guilin. Caves are rare in dolomite hills, closed depressions are poorly developed, but small springs occur along the borders of the dolomite.

The distribution of the residual Guilin peaks in the peak forest plains is thus not accidental or random, but depends upon the reaction of the rocks and their lithology to the agents of corrosion and erosion and their setting in the

Photo 8. Dolomite hills, near Ertang, Guangxi

structural context. Each peak can be explained as a result of its situation and its lithololgical and structural controls. This must also apply to the development of the underground drainage; if all the major caves and large conduits are in the sparry allochemical limestones, then it should be possible to reconstruct the phreatic drainage of the area in past periods from the fragmental remains of the present-day caves. Reconstructions such as this and the reconstruction of the history of the actual landforms is work for the future. The structural controls of the fenglin have been analysed by the Sino-French team in Guilin (Drogue and Bidaux 1992).

4.2.1 *Caves in the Fenglin*

The caves associated with the peak forest are more or less horizontal galleries accompanied by bigger chambers due to collapse (Photo 9). Phreatic tubes as in Reed Flute cave related to main epiphreatic water table caves are also common. There are, according to Yuan Daoxian (1980), three layers of caves in the Guilin area. (1) *first layer caves* (relative height ± 0m) occur near the present water level of the Lijiang, as at Elephant Trunk Hill cave; (2) *second layer caves* (relative height 7–15m) correspond to the first terrace.

Peak Forest (Fenglin)

Photo 9. Cave in fenglin, with solution slots, Guilin

These are quite common, though they are within the zone of seasonal fluctuations of the water and, therefore, have few stalagmites. The entrances to these caves are often at the bases of the peak forest hills – where they are today, influent caves. The scallop markings on both the roofs and sides of these caves always indicate that at the present-day water flows into the limestone hills from the plain (Zhu Xuewen 1988a); and (3) *third layer caves* have a relative height of 25–35 m. This group includes Seven Star cave, Reed Flute cave and Tunnel cave. Some of these caves are beautifully decorated with fine stalagmites up to 40–50 m high. It is possible that there are cave layers above these, but they are sparsely distributed and insufficiently studied (at 50, 90, 150 or even 300 m). There also are caves below the level of the Lijiang which have been encountered in drilling, down to depths of 80 m below the bed of the river (approx. 60 m a.s.l.). Fauna of early to middle Pleistocene containing *Gigantopithecus* occur in high level caves, usually over 80 m above the present water-level. Caves of the third layer in Guangxi may contain fossils of *Ailuropoda-stegadon* fauna – a good many of these animals are extinct or no longer live in Guangxi. In the Zhenpi cave 5–7 m above the river, 8 km S of Guilin city, skeletons of man of Neolithic age and modern fossils were found which have been dated to 7500–10 990 B.P. Caves below the level of 40 m above the river are middle to late Pleistocene in age (Yuan Daoxian 1980).

The two main tourist caves in the Guilin city area are Reed Flute cave and Seven Star cave. These make quite a contrast. Reed Flute cave is no longer associated with the water which might have formed it; it is basically an old paraphreatic cave. It is one of the best decorated in the region with 30–40 m high stalagmites. The main chamber is a great collapsed feature in sparry limestones. Seven Star cave is closely associated with a distributary of the Lijiang and has much less stalagmitic decoration. The modern river flows in galleries beneath the show cave. Seven Star cave (Qi Xing Tong) was already a tourist cave in 590 A.D. and was discussed in a book of the Tang Dynasty in 899 A.D. Water dripping through the joints in the limestones is often very undersaturated and corrosive, particularly if it has passed through loose Quaternary sediments. The average total dissolved solids content of the water ranges from 180 to 250 mg/L. The dripping water in the caves can be oversaturated with respect to CO_2 and the pCO_2 index can be 20 times more than that of ordinary karst water. Dripping water in the caves is particularly important in the rainy season (April to August), which is also the time of greatest carbonate deposition in the caves. The rate of growth of the stalactites and stalagmites varies between 0.01 to 3.8 mm/year. Condensation forms of stalactitic deposition, particularly shields (or palettes) are well represented in Reed Flute and other caves. At the Lotus Throne cave near Xingping, stalagmitic deposits formed under water (in a lake) and by water dripping from the roof, are some of the most remarkable in the whole area (Photo 10).

Photo 10. Throne-like formations in Lotus cave, near Yangshuo

These deposits are no longer forming – the lake having become drained (Zhu Xuewen and Zhu Dehao 1988).

A great deal of data have been collected by Zhu Xuewen and Zhu Dehao (1988). They summarize work on ^{14}C and U/Th dates on the stalagmites and stalactites, while attempts have been made at dating by ESR (Williams et al. 1986). Results show that the age of the oldest stalagmites is over 600 000 B.P. The ^{14}C results indicate that stalagmite deposition continued throughout the last 40 000 years, though there is a hint of a gap between 20 000–10 000 B.P. The U/Th results are less reliable, as the amount of uranium in the stalagmitic calcite is very low and specimens are often contaminated. A group of stalagmites from Da yan cave – Maomaotou cave dated by Angela Rae in 1983 give dates ranging from 385 000 ± 152 to 191 000 ± 24 with a hiatus at about 220 000 ± 30/25. The hiatus is marked by a layer of hydroxyapatite – $Ca_5(PO_4)_3(OH)$, suggesting a dry phase when $CaCO_3$ deposition did not occur. Another hiatus in deposition occurs around 190 000 B.P. Thus, $CaCO_3$ deposition and stalagmitic growth have been episodic presumably caused by the Pleistocene climatic fluctuations. Wang Xunyi et al. feature a stalagmitic column 97 cm long, which shows moderately rapid growth up to about 267 000 B.P.; then slow growth up to 174 000 years; and followed by rapid growth again after 170 000 B.P. (Wang Xunyi 1986). This clearly reflects the possibly cooler and drier phases Guilin would have experienced in the Pleistocene and also its warmer and wetter phases. These data also seem to fit the Quaternary chronology derived from the loess palaeosol studies (Liu Tungseng 1988). Zhu Xuewen (1988a) shows a reconstruction of the Pleistocene events in the Guilin area. The effects of these variations on the karst landforms are still unknown.

Zhu Xuewen (1988a) has further information on some of the most important caves in the Guilin-Yangshuo area including analyses of the rocks in which the caves occur; heights and lengths of cave passages; water analyses; measurements of pCO_2 in the caves (the last two with their seasonal variations); and stable isotope data. Meteorological measurements were made in four of the caves, including the Maomaotou cave and Tunnel cave, for about 1–1.5 years; cave humidity and temperature were both measured. The results showed that the larger caves have a distinct meteorological zoning, with rather lower temperatures than the average (19.2–21.5 °C) and less variation; the smaller caves showed more variation. The caves are affected by the meteorological conditions outside but with a time lag due to the thickness of the cave wall; the temperatures are lower with a thick wall, but the mean humidity higher. It will be seen that this book contains some of the most detailed work on caves anywhere.

Non-carbonate deposits are also important in the caves, gravels, sands, clays and muds. Many of the muds have been removed by farmers to spread on the land, so sequences of clastic deposits are not easily found. Some of the deposits in the higher caves have become lateritized and are reddish/brown in colour; in the lower caves, the clastic deposits can be yellowish or grey. However, the works of Williams (1987) and Bull et al. (1989) have shown that the

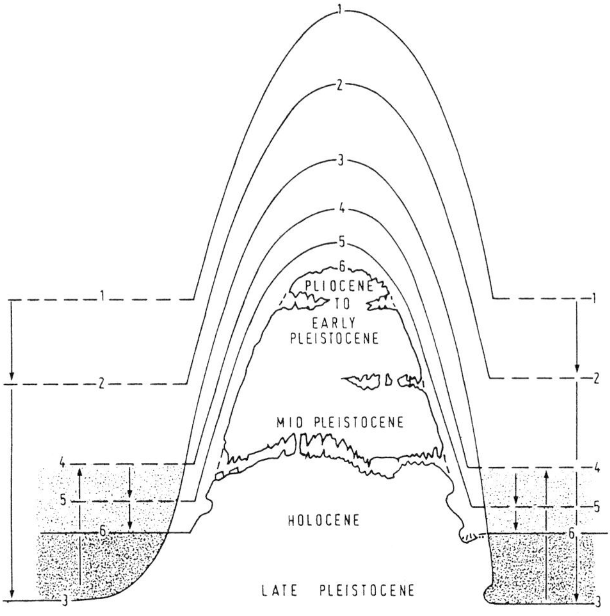

Fig. 20. Tower karst evolution near Guilin. *Numbers* indicate the schematic stages from *1–6*. (Williams 1986)

deposits in the higher caves are not always older than those occurring in the lower level caves and the sequence in the caves is probably quite complex (Fig. 20). Mudflow deposits, for instance, occur at many different cave levels. Williams et al. have examined the palaeomagnetic record of cave sediments and speleothems from Nan cave in the Tunnel hill groups of caves (Williams 1987). The sediments investigated are in a cave 23 m above the Lijiang. Their palaeomagnetic record suggests a reversed inclination and Williams concludes that the sediments may be 900 000 years old and may even be as old as 1 600 000 years. The significance of these dates and the ages of the stalagmites will be discussed later. Some new work by Lin Jinrong (1992) suggests that Triassic deposits occur in some of the caves, which if correct indicates an even older origin for the Guilin caves.

4.2.2 Hydrology of the Fenglin

The hydrology of the peak forest plain is of the aquifer type though the flow is by no means homogeneous. The karst aquifers depend upon the degree of karst development rather than the porosity of the limestones. Artifically induced collapses occur in the limestones and in overlying Quaternary and Holocene deposits near Guilin, often because of overpumping of the aquifer (Fig. 21). Natural collapses or blue holes (at the level of the water table) occur on the peak forest plain, south of the road to Yangti; these are often on the

Peak Forest (Fenglin) 77

edge of the peak forest and the peak cluster and indicate an inter-connecting water table, a few metres below the limestone surface. They form a group of holes near the water table and form small lakes up to 5–6 m deep (Photo 11). Underground drainage of the peak forest aquifer can be intermittent, as streams sink into influent caves at the foot of the residual peaks and then emerge from them as springs. Horizontal flood and swamp corrosion notches are quite common where such streams are dammed into lakes by debris and vegetation or by floods in the wet season. Probably such corrosion notches represent local conditions rather than any paticular regional flood levels or events (Photo 9).

Beneath the surface, in the aquifer itself, flow can be quite rapid in distinct underground streams which have been located by drilling bore-holes. These streams are in the epiphreatic caves being formed today, which when the area is uplifted will become drained and accessible as the horizontal caves seen in the peak forest hills today. As Zhang says, "we are ... fully justified in thinking that the well-galleried horizontal cave system seen in the peak forest hills today was moulded by the pre-existing underground river system near the water-table" (Zhang Zhigan 1980a). In the Guilin peak forest area, however, these galleried horizontal cave systems have suffered a great deal of dissection as the karst was eroded and now only remain as conspicuous fragments in the tower karst residuals or as high level arch-like relics, as at Moon cave near

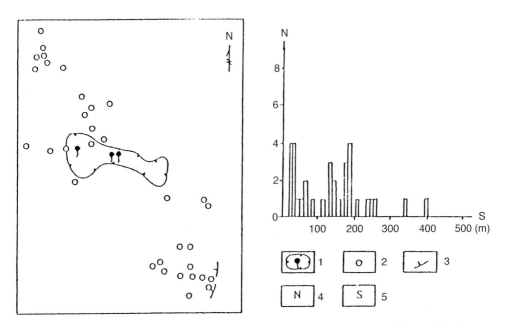

Fig. 21. Distribution of collapses following pumping from Shanyanjing Bluehole, Guilin, Guangxi. *1* The Shanyanjing bluehole; *2* collapse; *3* soil fissure; *4* number of collapses; *5* distance of collapse point from Bluehole

Photo 11. Water-table "blue hole" near Putao

Yangshuo, or in the arch above Nanxu, and in Tunnel cave near Guilin (Photo 12). The external form of peaks like Tunnel hill is thus clearly younger than the caves which cut through them. As already mentioned, the caves towards the foot of the hills are in general younger than those towards the top, but relative elevation is not a sure guide to their age. There is evidence, for instance, that aggradation of the Lijiang prior to 34 000 B.P. could have buried the Chuangshan cave in the Tunnel Hill and filled it with sediments (Williams 1987, Bull et al. 1989). These sediments were later washed out and re-excavated by the river. The caves have also been modified by the inflow of waters from the plain (Fig. 20).

The karst towers, therefore, grow by the lowering of the bedrock floor around their bases, but this lowering has been intermittent in the Quaternary and the Holocene and interrupted by phases of aggradation. Williams (1987) calls the peak forest towers, *time-transgressive landforms*, as they are older near their summits than at their foot. In order to maintain the form of the towers in the Guangxi peak forest plain, the rate of uplift of the earth's crust has been largely equal to or slightly less than the general rate of karst denudation (calculated as 89.68 mm/1000 years) (Zhang Zhigan 1980a). The landform

Photo 12. Contact between limestones and non-limestones (foreground) near Nanxu and collapsed cave arch (fengcong)

development of the peak forest plain has proceeded continuously through time, with the valleys generally being deepened (with interruptions) and the peaks growing relatively in height with the slow uplift and lowering of the bedrock floor. Zhang believes that the amplitude of each uplift has not been sufficient to alter significantly the original pattern of the hydrographic network (1980a).

4.3 The Peak Cluster (Fengcong)

Clusters of steep-sided peaks, generally form the "high country" of the Guilin karst. The slopes of the even-sided peaks vary according to the lithology that has already been described and can be 30°–40° in less strong limestones and 60°–70° or even more in the strong limestones. However, slopes are usually the same round each of the peaks. The clusters are interrupted by outcrops, usually N-S of the impermeable sandstones (see map, Fig. 16), and the sandstones are important in controlling the distribution of the sink holes, springs, depressions and valleys. Peak cluster is particularly important in the area traversed by the Lijiang gorge and the area to the east of the periphery of the syncline and where recent uplift has taken place – along the Haiyangshan faulted area (Photo 6). The highest limestone peaks about 700 m a.s.l., are often capped by insoluble residuals of sandstone; peaks on the flanks of the syncline are higher everywhere. Yuan Daoxian (1980) suggests that there are various concordant peak levels, at about 70, 150 and 300 m above the Lijiang water level; but so far no detailed work on peak summits has yet been done. The individual hills of the clusters vary greatly in height – some being only 20–30 m high, others 300–500 m; and in diameter also smaller hills are only a few tens of metres in size, whereas the larger ones may be 1–2 km in diameter. The limestone peaks often "sit" on sandstone hills as conspicuous residuals towards the edge of the basin.

Depressions of various sizes occur between the peaks. These may be small (a few tens of metres in diameter and depth) and are closed depressions or dolines; or they may be longer, narrower depressions, up to a few kilometres and were formerly fluvial valleys, but which are now dry, and wind between the peaks. The floors of such former fluvial valleys are intersected by the smaller closed depressions, so that they now have an undulating karstic profile. Such a dry valley is seen to the ESE of Yangti – east of the Lijiang valley (Photo 13). The floors of the smaller dolines lie above the flood season water table (i.e. are in the aeration zone); they usually contain swallow holes or sink holes, which are temporary inputs for water. Their slopes usually maintain the angle of slope of the peaks above them as they are fashioned by the same slope-solution processes which are determining the peak slopes. Their bases, however, have become steepened or become puddled with debris, mud and soil, so that they level off and may form small areas of relatively flat land, enough to support a field or two for cultivation. The slopes of the dry valleys

Photo 13. Dry valley in fengcong (closed depression in cockpit karst), east of Yangti

are also as equally steep as the peaks above. Their bases are more varied and parts may be seasonally flooded and lie in the zone of variation of the seasonal water table. The parts not inundated become karstic cols and interrupt the valleys. Intermittent springs give rise to small streams which later sink into swallow holes; both may function as estavellen (Zhang Zhigan 1980a). If the area between the peaks is floored by the clastic rocks, it may be drained by a perennial stream; this may be the beginning of more widespread alluviation to form a plain. Such perennial streams are rare in the peak cluster of the Guilin area. If sufficient inputs for water develop depending upon the lithology and degree of jointing in the bases of the dolines and dry valleys, a polygonal karst hydrological network may be set up, as described by Williams (1972); this is discussed in relation to the area west of Yangti by Zhu Xuewen Zhu Dehao (1988, p. 29) and Zhu Dehao (1982; Fig. 22).

Zhu Xuewen and Zhu Dehao (1988) examined the morphometry of the fengcong (peak cluster) areas, particularly in an area west of the Lijiang, near Yangshuo. They have used the methods developed by Williams in his studies of polygonal karsts in New Guinea (Williams 1972). Over 8000 measurements were obtained. In the area examined, polygonal depressions (in the William's sense) account for 62% of the total surface – the average number of sides according to Zhu Dehao being 5.3. He claims that the depressions tend to be evenly distributed in plan (nearest neighbour R = 1.64). The following relationships also occur: (1) the smaller the area of the depression, the higher the mean elevation of the bottom of the depression and vice versa (Fig. 23); (2) the smaller the area of the depression, the smaller the height difference between the depression bottom and the peak summit; where the area is greater, the height difference is greater; and (3) the altitudes of the bottoms of the depression above sea level and the peaks have a linear but inverse proportional relationship, i.e. the higher the elevation of the bottom of the depression the smaller the height difference detween the depression and the peaks (and vice versa). These regularities are similar to those discussed by Williams.

Further morphometric data are given by Yuan Daoxian et al. (1991), who indicates that polygonal cluster depressions stretch along the slope zone between the Guizhou-Yunnan plateau and W Guangxi for 1000 km. These facts suggest that the lowering of the peak summits is slower than that of the downward deepening of the depressions; this is due to the accumulation of soil and vegetation in the bases of the depressions causing strong biogenetic corrosion. Thus, as Zhu Xuewen (1988a); stated, the positive topography (i.e. the peaks) is always "older" than the negative topography (i.e. the depressions); as a result, the tops of the peaks may be preserved in the landscape for a long period of time. The development of the depressions shows a tendency of continual subsidence at the bottom and increasing polygonal area. As the depressions deepen and enlarge, their evolution slows down. The polygonal network concept is useful, but it may sometimes be more apparent than real; unless accurate water tracing has been done, it is difficult to know precisely what the underground drainage is really doing.

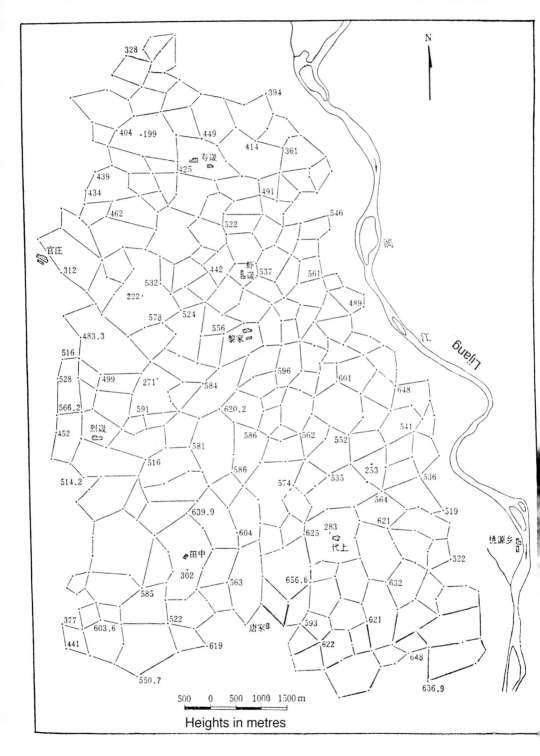

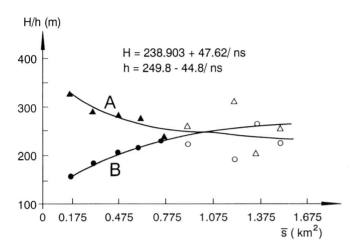

Fig. 23. Relationship between area of depression and height in area W of the Lijiang. (Zhu Dehao 1982)

A well-developed vadose and aerated zone characterizes the hydrology of the peak cluster. Vertical infiltration and penetration of rainfall takes place and vertical dissolution predominates. The caves are developed in the thick vadose zone, well above the water table. Laterally developed caves are rare, but dendritic cave passage forms do occur. The large collapsed type of cave associated with the peak forest is exceptional. The cross sections of the caves are rectangular (not big phreatic tunnels) and are of the fissure type. Vertical caves occur in the depressions and are formed by seepage or small collapses, related to the vertical dissolution; they are also related to vertical linear water flow coming off the sandstone and clastic mountains which inter-digitate with the limestones of the peak cluster. Such water from the sandstones can be quite aggressive – Zhou Shiying et al. calculate the karst denudation rate for two peak cluster depressions in the Guilin area to be 89.68 mm/1000 years (Zhou Shiying et al. 1988).

The most impressive aspect of the peak clusters is the generally steep slopes often greater than 60° particularly in the strong, sparry limestones of the Rongxian formation. It is also possible even in the polygonal karst area studied by Zhu Xuewen to detect many of the original lines of drainage. Slope forming processes have been observed in the depressions of the Guilin Experimental Site "where every patch of soil on the doline slope has a perched spring" and "the discharge may be small but quite a few springs flow the whole year round, and survive the dry season" (Yuan Daoxian and Drogue 1988). Thus, fluvial slope processes are important even today, particularly in the epikarst layer and are continuing a process begun as soon as the limestones were exposed and uplifted. Work at the Guilin Experimental Site has revealed the importance of the epikarstic layer which varies from 3–10 m in depth. Because of its connection with the soil and plant layers, the CO_2 in the soil air

Fig. 22. Polygonal network of closed depressions, west of the Lijiang, near Yangshuo. (Zhu Dehao 1982)

of the epikarst is 10–100 times higher than the CO_2 in the atmosphere. Measurements at the Experimental Site showed that in winter the soil air content was about 0.07% atm in January but that in July in Summer it was 2% atm. Water infiltrating into the epikarstic layer is quite aggressive and the hardness of the fissure water is 170–200 mg/l. However, springs emerging from the epikarstic zone often lose CO_2 and precipitate tufa and the average carbonate hardness of the springs at their source is about 150–160 mg/l (Yuan Daoxian and Drogue 1988). Slope processes in the peak cluster would justify further study and help to explain the steep and conical nature of the peaks.

The steep slopes of the peak cluster are likely to have formed by strong uplift; at first, the fissures in the limestones would not be able to absorb much of the rainfall and in the first stages of the peak development rapid runoff by means of gullies would be much more important than now. These gullies would be steep – due to the original uplift and the intensive and heavy runoff from the rainfall. Between such gullies, steep-sided peaks would be left. The gullies would feed larger fluvial valleys below the peaks – the origin of the present network of dry valleys which can now be seen in the peak cluster. As fissures become widened in the rocks, more of the water in the gullies would become absorbed, except in heavy storms. Less water from the gullies would reach the valleys. Eventually, input points for underground water develop in both the gullies and the valleys, and closed depressions begin to originate. Meanwhile, because of the rapid runoff, the peaks – contributing water to the gullies and dry valleys remain steep, indicating the influence of both rapid uplift and the rapid and torrential runoff caused by heavy monsoonal or tropical rainfall. In this explanation, the steep peaks of the peak cluster are interpreted as being formed partly by normal fluvial processes, superseded in time by the depressions developed by karstic processes. In areas of strong tectonic uplift and rapid runoff, the heritage of original fluvial drainage pattern may be detected for a long time in the landscape. The watersheds between the original gullies and valleys remain in the landscape as the peaks, while the gullies and valleys become karstified. The peaks are broken down by a combination of processes, including the slope process indicated above (which tend to act more or less uniformly all round, so as to give the conical shape) and also by lateral solution processes which wear back the slopes at the foot of cliffs as described by Bremer in her work in Sri Lanka. These latter processes give rise to sharpened arête-like peaks which are a feature of the Lijiang gorge between Daxu and Yangti. These arêtes have not been formed by the coalescence of closed depressions, but basically by processes of lateral solution and pedimentation (Bremer 1981; Sweeting 1990).

4.3.1 Caves in the Peak Cluster

Detailed information about the caves and underground drainage of the peak cluster area, east of the Lijiang, was collected by the joint Anglo-Chinese

The Peak Cluster (Fengcong)

project on the caves of S China (Fig. 24). Here, thicknesses of over 3000 m of upper middle Devonian to the Lower Carboniferous carbonate rocks occur (Eavis and Waltham 1986). One of the most important caves mapped was the Guanyan cave system which extends from the small polje of Nanxu in the east, to the Lijiang river, 7 km away. Not all of the 7 km are accessible. The river Chuan Yan which drains the Nanxu polje sinks into the limestones in a fine cave passage in a wall of rock at least 200 m high (Photo 12); this cliff has been produced by the recesssion (by collapse mainly) of the cliff-like peak cluster and peak forest from the sandstone hills to the east (2–3 km distant). Near the sink of the Chuan Yan in the cliff is a rock arch which is possibly the remains of a former high-level cave once part of the system. The Nanxu river probably sank into the limestones soon after the Tertiary uplift. The Chuan Yan part of the cave can be followed for over 3 km. This cave, like others in the area, is simple and relatively linear without the development of extensive networks. In the floods of the wet season, large amounts of coarse gravel are transported through the system and help to explain the size of the caverns and the collapse features (as well as a result of the lithology). In Guan Yan, remnants of earlier gravels cemented by stalagmite can be found adhering to the walls and at some points to the roof of the passage (Smart et al. 1986). The cave of Chuan Yan ends in a sump after 3860 m of passage. There are several indications of former high level routes and the cave intersects at least two deep dolines (Fig. 24). The cave is then seen again in Xiaoheli Yan which is 2840 m long. Below Xiaoheli Yan is the Xiaoheli window, which is a large doline in the karst with the Guanyan river crossing its alluviated floor; this large depression receives water from a sandstone outcrop in the NW (Fig. 24). The final part of the Guanyan cave system is Guan Yan (Crown cave) itself, a 3830 m passage and 65 m high in parts – which gives a through route from the Xiaoheli depression to the Lijiang, where there is a fine resurgence into the river. It also has a high level route. At the resurgence the flow ranges from less than 300 l/s in the dry season to a wet season flow of over 8000 l/s. The catchment is over 80 km^2.

Evidence of earlier surface water courses crossing the peak cluster between Nanxu and the resurgence of Guan Yan can be seen in linear depressions developed on the valley floor – well seen between Nanxu and the Xi Tang lakes. High-level gravel trains also occur. At Maliu Kang, water-worn cobbles "form an extensive spread at elevations up to 150 m above the persent river sink hole" (Eavis and Waltham 1986).

In the upstream chambers of the Xiaoheli Yan there is ponding behind sediment-choked collapses. Smart et al. (1985b) is of the opinion that ponding behind the high-level overflow in Xiaoheli Yan may be responsible for the survival of the Xi Tang lakes, which appear as an anomaly in the middle of the karst; in addition, the lakes receive drainage from shales and sandstones intercalated with the limestones of the Yanguan stage. The water level of the lakes fluctuates more than 11 m between the wet and dry seasons.

Note that this cave system cuts across the geological grain of the country and does not even seem to be greatly affected by the strike or tensional or shear faults. This suggests that the system has been formed rapidly in response

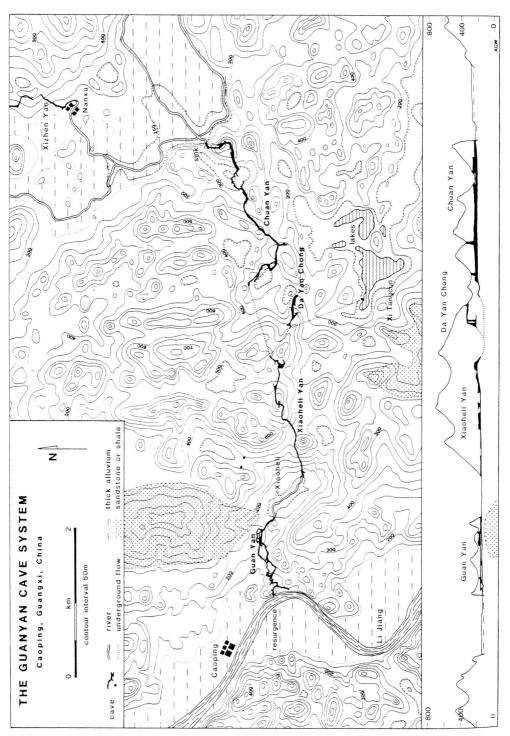

Fig. 24. The Guanyan cave system

to the cutting down of the Lijiang in its gorge. It would be expected that caves in this area would be influenced by the faulting or by the generally N-S strike of the rocks, but the rapid (presumably) development of the hydraulic gradient to the Lijiang has had the most powerful effect upon the underground drainage. This may be why in this location the longest cave system in the Guilin area has developed. The same effect is seen in the peak cluster area west of Yangti, where a stream which drains the west to east trending Sihe polje formed in Devonian sandstone, sinks about 70 m above the river into the limestones to the east end and breaks through the Lijiang cliffs to emerge as a resurgence above the river on the western bank of the river. The development of the Lijiang gorge has prevented adjustments to structure of the underground drainage in the areas of peak cluster close to it. Further away – from the Lijiang, nearer to the Haiyang Shan – some of the caves mapped by the China Caves Group seem to have N-S trends and to follow the geological structures more closely.

4.4 Age and Origin of the Guilin Karst

Basically, the Guilin karst is of two types: first, the limestone residual towers which form the peak forest plain and largely concentrated in the west of the area from Guilin southwards; and, secondly, the peak cluster or elevated peaks which form most of the area on either side of the Lijiang from Daxu to Yangshuo. There are some interdigitations and mixing of the types due to local conditions, but this is a broad distinction.

It is clear that the residual towers of the peak forest plain are very old landforms. Williams et al. (1986) have recently examined cave sediments and speleothems from Nan cave in the Tunnel hill Group. They find that the palaeomagnetic record suggests a reversed inclination and that the sediments may be 900 000 or even 1 600 000 B.P. These sediments are in a cave, 23 m above the Lijiang. As we have seen, the stalagmitic records from the caves in the peak forest go back probably at least 600 000 B.P. The caves themselves are clearly very much older. What we are dealing with in the peak forest area of Guilin are the relics of an ancient karst landscape. Beneath the peak forest plain at Guilin there are at least 600 m of Devonian limestones in the axis of the syncline; if the area were to be rapidly uplifted today, then this ancient landscape could be rejuvenated – the towers surviving as residuals on a plain or plateau surface waiting to be dissected. Drogue and Bideaux give a tectonic explanation in the region of the Guilin Experimental Site (1992).

It is the peak cluster which poses problems. Lin Yushi et al. (written in 1984 in Chinese, quoted in Yuan Daoxian 1985b) have studied in detail the Cretaceous red beds which are visible in different parts of the Guilin karst and best exposed at Guilin airport between Qifeng and Yanshan, where the beds are about 100 m thick. The beds consist of red mudstone, siltstone and conglomerate, but closely associated with them is a basal red breccia. The red breccia is believed to be late Cretaceous because of its occasional fossils of *Atopochara* flora sp. *Charites*. Detailed examination of both the peaks of the peak forest and the peak cluster has shown that there are widespread occurrences of the breccia in many parts of the Guilin karst – at many levels as high as from 580 m on Houshan hill to the peak forest plain itself. The breccia is composed of carbonate gravel, often cemented with reddish argillaceous mud – sorting and lamination can be seen. This cemented breccia occurs on many of the hills and in caves particularly in the area west of the Lijiang, but is also seen on peaks and in depressions east of Yangti. The conclusion is, therefore, that these late Cretaceous deposits covered at one time all the limestones of the Guilin region up to at least about what is now a present level of about 600 m. The deposits near the airport form the largest area of red beds remaining, which protect the limestones from subaerial denudation. The Cretaceous red beds have been removed by erosion as a result of upwarping (possibly enhancing the original Yan Shanian folding) at some time in the Tertiary. A simplified model for the evolution of the Guilin karst is given by Yuan Daoxian (1985b; Fig. 25). Evidence suggests that the upwarping in the Guilin area has been asymmetrical, with the greatest amount of uplift east of the syncline, the main area of the peak cluster, and relatively little upwarping in the west, the main area of the peak forest plain. Yuan Daoxian thinks that the period of greatest uplift and the greatest removal of the Cretaceous red beds was in the Quaternary, due to the high rate of corrosion (100–300 mm/1000 years) of the Guilin limestones. If the limestones had been largely exposed in the early Tertiary, his argument is that *all* traces of the red beds and breccias would have long since disappeared. The total Quaternary denudation is believed to be of the order of 100–150 mm. Yuan Daoxian does, however, make the point that different parts of the limestones would have been exposed at different times so that the evolutionary history of different parts of the Guilin area would not necessarily be the same. This idea is confirmed by the concepts of some structual geologists who believe that Guilin area is made up of several structural blocks which have reacted separately to the tectonic forces. Ideas along these lines were also suggested by Wang Nailiang in the 1960s.

Further study on the age of the Guilin karst has been contributed by Lin Jinrong (1992). He claims that caves have been found which contain Triassic sediments, citing caves near the Changhai Machinery Factory (near Guilin). These sediments are lake basin deposits and resemble to some extent the later Cretaceous deposits already discussed. Lin Jinrong (1992) implies, therefore, that the Guilin limestones were already karstified as early as the Triassic period. He gives maps of both Triassic sediments and of the Upper Cretaceous sediments remaining upon the limestones of the Guilin area (Fig. 26). He suggests that the Cretaceous beds were deposited in an inland lake basin and that 500–600 m of clastic sediments accumulated. Solution by lake waters

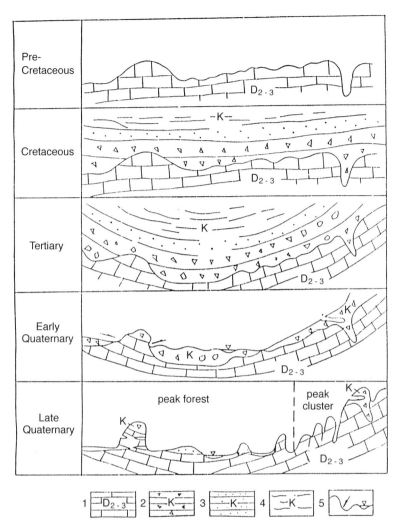

Fig. 25. A Simplified model for the evolutionary history of Guilin karst. *1* Middle and upper Devonian limestone; *2* Cretaceous red breccia; *3* Cretaceous red siltstone; *4* Cretaceous red mudstone; *5* surface stream and swallet (Yuan Daoxian 1985)

might also be responsible for some of the cave levels. Removal of these Cretaceous beds took place during the Tertiary and Quaternary as discussed above by Yuan Daoxian (1985b). Lin contends that the uplift by the east-west trending Nanlin mountains to the north of the Guilin area in the late Tertiary or early Quaternary may have had an important effect on the Quaternary climate of Guilin; the mountains would have formed a block to the movement of cold air from the north and warm air from the south. The resulting high rainfall and storms would have contributed to the landslides and debris flows

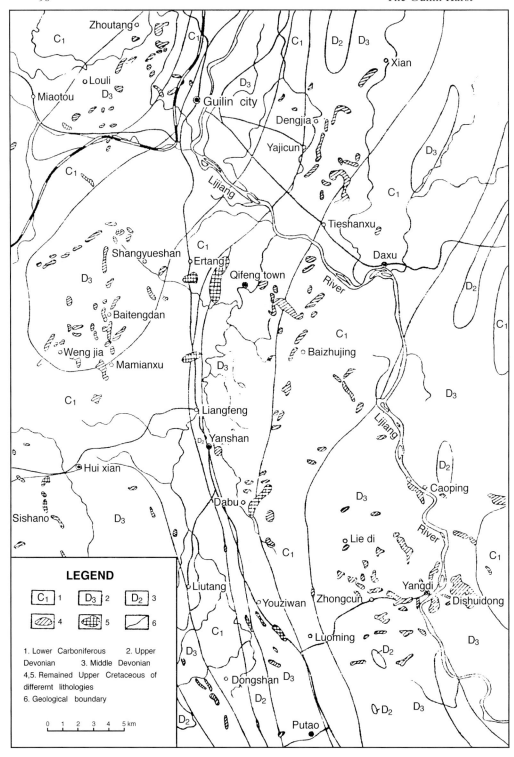

in both the carbonate and non-carbonate landscapes of the Guilin area (Lin Jinrong 1992).

We do not know what the surfaces of the limestones were like when the Cretaceous beds were laid down, though as we have seen, they had already been karstified to some extent. It is also difficult to assess the amount of corrosion of the limestones that took place beneath the Cretaceous red beds as subjacent solution of limestones can sometimes be important and a crypto-karst formed (Nicod 1992). The state of the surface of the limestones on the exposure and removal of the red beds are, therefore, not known. Once exposed, modern fluvial and karst processes would take over. Rapid runoff and gullying under a torrential monsoon rainfall regime would produce an original drainage system upon the limestones and the steep slopes of the peak cluster hills. Later infiltration into the limestones produced the karstic dry valleys and depressions and the underground drainage formed the caves.

It will be seen that an explanation along these lines helps to account for the general distribution of the peak forest plain and the peak cluster, and for the relatively recent uplift in the eastern part of the area. This accounts for the incision of the Lijiang, as it was a meandering river, possibly on the red beds during the early Quaternary or late Tertiary (Lin Jinrong 1992). It also explains the steep slopes of the peak cluster. Such differential uplifts led to the differential landscapes in the Guilin area. This interpretation of the geomorphology is supported by an analysis of the drainage patterns in the Guilin area. The upper parts of the streams flow generally towards the N but their lower courses flow predominantly southwards, suggesting a more recent rearrangement of the drainage pattern. P.A. Smart (in Eavis and Waltham 1986) suggests that not only were the uplifts greater in the eastern part of the Guilin area, but that the uplift plunged towards the SW. This might explain the later southerly reversal of the drainage.

On this interpretation, the Guilin karst did not begin to take shape until the removal of the Cretaceous beds. If this removal was largely Quaternary, then the Guilin landforms are not as old as has often been thought. I believe that the middle and later Tertiary must also have seen erosion of the red beds and the origin of most of the karst landforms. The nature of the Cretaceous (red bed) surface is also unknown – and to what extent the karst is made up of relict landforms. However, corrosion and erosion of the limestones during the later Tertiary and even during the Quaternary are likely to have been substantial, so that relatively few Cretaceous and early Tertiary remnants could have survived.

The zonation of the karst landscapes so often described in Chinese literature on S China is not very obvious in the dissected karst of Guilin. However,

◂─────────────────────────────────────

Fig. 26. Remnants of upper Cretaceous sediments in the Guilin-Putao area. (Lin Jinrong 1992)

in other areas of Guangxi, it is assumed that from the lower to the upper reaches of a river, the zones of relief are approximately in the following order: peak forest (fenglin) plain, peak cluster (fengcong) and valley, and peak cluster (fengcong) and depression. It may be that further detailed work such as has been done in Guilin may modify these ideas of evolutionary sequences, but as will be seen in the discussions which follow, evolutionary sequences are still part of Chinese thinking on the development of many of the karst areas, as distinct from some of the newer ideas of the workers in Guilin.

5 The Cone Karsts of Guizhou

Guizhou province covers 176 000 km² of which 72% is karstic. The Guizhou plateau has an altitude of between 1000 m a.s.l. to over 3000 m, but is mostly between 1000–2000 m. The plateau extends into the neighbouring provinces of Yunnan, Hunan, Hebei and Sichuan (Fig. 2). It is part of the second great topographic step of China and falls to the E to neighbouring Guangxi. The plateau slopes generally from W to E. It is drained to the N by major tributaries of the Changjiang (Yangtse) and to the S to the Zhujiang (Pearl) river. The average annual rainfall is about 1100 mm, mostly coming in summer with considerable winter drizzle on the plateau. Guizhou has a cloudy climate, like that of Sichuan, as a result of the meeting of different air masses which arrive from the NE in winter and the S in summer. Average temperatures are 11–19 °C. Mountain ranges run generally NE-SW, except the Miao range which is E-W.

Almost all the rocks of Guizhou are sedimentary, ranging in age from the Proterozoic to the Quaternary. Up to the Triassic these rocks are mainly shallow sea marine in facies and dominated by limestones 8500–12 000 m thick. The limestones which support the karst features are chiefly the Carboniferous, Permian and Triassic, but some Cambrian limestones are locally important. Structurally, the area is quite complex. The main folding dates back to the Indosinian and Yanshanian movements (Triassic and later Mesozoic). In NE Guizhou, NW Hunan, and W Hubei, the NE-SW tending folds are high angled and close, and part of the S China fold system (Fig. 3). These give rise to long, narrow and steep limestone outcrops – linear karst belts. In middle and western Guizhou the folding is less intense, and dips are generally lower (though occasionally vertical), and thus a greater area of limestone is exposed. Several intercalated non-carbonate strata occur in this area. All these strata in Guizhou were much affected by the Himalayan earth movements; Guizhou lies between the area of intense uplift in Qinghai-Tibet and the area of more gentle uplift in Guangxi. It is believed that Guizhou was uplifted by about 1000–2000 m in the later Tertiary and that even the amplitude of the Quaternary uplift was from 300–500 m. The structures in Guizhou are NNE-SSW and NE-SW in the north and east; N-S in South Guizhou and NW-SE in western Guizhou. In the extreme W of Guizhou and in the neighbouring E Yunnan, Tertiary and Quaternary movements produced N-S block faulting (Yang Zunyi et al. 1986).

The most important karstic limestones are the Baizho dolomites and the Datang and Maping limestones within the Carboniferous. Each of these formations is several 100 m thick and occur in particular in the Shuicheng area in W Guizhou. In the Permian, the Maokou and the Qixia limestones are very important. In the Triassic, the Middle Triassic Falang limestones are a massive bioclastic series, over 300 m thick and form the basis of the famous Longgong karst. The Lower Triassic Guanling limestones and dolomites are represented in the Huangguoshu area. The carbonate strata alternate with clastic rocks, giving rise to *multi-layer* water-bearing layers and karst belts. Frequent movements during the deposition of the limestones gave rise to a series of planes of unconformity. There are six planes of unconformity in Guizhou – these become the later planes of karst denudation when karst erosion reaches individual base levels. The karst rocks are only sparsely covered by Mesozoic shales and sandstones and some Tertiary red beds and Quaternary deposits – so Guizhou is essentially an area of bare karst.

Strong uplifts in the Tertiary and Quaternary considerably exceeded the rate of karst denudation. Down-cutting did not merely deepen the orginial valleys, but gave rise to the canyon systems; this development has destroyed the original karst hydrographic networks. A new phase of karstification was initiated, particularly by headward erosion and corrosion which has helped to break up the pre-existing planation surfaces; the karst processes have been revitalized with the canyons as the base level for discharge (Fig. 27). Away from down-cutting rivers and the canyons, it is believed that the pre-existing planation surfaces are preserved and the older karst networks have continued to develop – the ancient hydrographic networks here acting as the base levels for discharge. The landscape of the Guizhou area is thus made up of two highly contrasting units within a unified karst region – the plateau areas and the canyon areas – the two units differing greatly in their landforms and in their underground hydrographic networks. It will be seen that this interpretation of

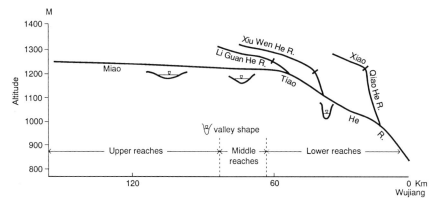

Fig. 27. Longitudinal profile of the Maotiao river and its major tributaries, Guizihou plateau. (He Caihua 1987)

the landscape units of Guizhou resembles a Davisian model and is like the explanations once given by H. Baulig for the landforms of the Grands Causses in the Massif Central of France, which was regarded by him as an old planed plateau surface area, strongly uplifted and now deeply dissected by the gorges of the Tarn and its tributaries. In many ways, though the relief is on smaller scale, and climatic conditions have been different, the Causses of the Massif Central have some analogies to the karsts of central Guizhou (Baulig 1934). The Dinaric karst in former Jugoslavia developed after the Miocene alpine folding and the later Tertiary uplift has developed in a somewhat similar way but is much more block-faulted and more evolved than Guizhou.

Those areas which have received particular attention in Guizhou are the Anshun-Huangguoshu karst in central Guizhou, and the Shuicheng area in W Guizhou. The characteristics of the plateau and canyon areas will be discussed, using these areas and others to illustrate the karst geomorphology. The area N of Anshun, around Puding, has been much studied by the Geography Department of Nanjing University; the town of Puding also has its own karst research station. West and southwest Guizhou have been studied by the Franco-Chinese speleological group, particularly in the Gebi(he) river area (Barbary et al. 1991).

The Guizhou plateau is uneven with an order of relief of 200–300 m. The railway line from Guiyang to Anshun runs along the plateau surface. The areas of more gentle relief are unlikely to be all of the same age or to be one major erosion surface now being dissected. Instead, each area of plateau surface is likely to have been "graded" to a series of local base levels – possibly within a specific time period in the Tertiary period or even at different times. Such local base levels are provided by outcrops of impervious strata in the limestones by the old planes of unconformity in the limestones, or by the formation of local water tables in the limestones caused either by local flooding or by alluviation of the karst plains and basins. There is a limit to the vertical drainage in these limestones possibly about 300 m, unless the relief is rejuvenated, i.e. the depressions are to a certain extent self-limiting. This is very clearly seen in the relief between Guiyang and Anshun. The hydrographic network on the local plateaux under these conditions is "mature". The aquifer is of a highly transmissive fissure flow type. In the larger valleys the groundwater table lies not far below the surface (from a few to 10 m). Where there is much alluviation and the sink holes are blocked, water flows on the alluviated surface in rivers or there may even be permanent lakes, for instance at Choahai lake at Weining in W Guizhou. Dolines, ponors and shafts are generally shallow and collapse of shallow caves in quite common. The rich caves form stalagmite and stalactite deposits, as in the caves near the city of Guiyang indicate a destruction and infilling of the former caves. Though modern karst processes are relatively weak the degree of karstification of the rocks is quite high, i.e. the rocks have many solutionally produced holes. The depth of solution beneath the water table may reach as much as 150–400 m. In the dry season, this zone is reduced to a few tens of metres.

The carbonate strata alternate with clastic rocks on the plateau, giving rise to karstic development in belts – in folded and fractured belts. The character of the plateau is diversified by the numerous cone-like hills which form a feature of Guizhou (Photo 14). These are more often isolated cones, i.e. fenglin, but groups of conical hills occur to form fengcong. The steepness of the cones is related to the lithology of the rocks; outside Guiyang city, between it and the airport, are low more or less rounded hills, with slope angles of 10°–15°; these are formed of Triassic thin-bedded limestones with much intercalation of shales and muddy layers. Hills of hard dolomitic limestone are common near Anshun and have steeper slopes, from 30°–40°. Very strong and well-jointed limestones may give conical peaks with slopes of over 60° as in the Longgong karst (Fig. 28). The cones on the plateau are characterized by their symmetry – the slope angles on all sides being approximately the same, irrespective of structure and distance from base level. The fenglin of the Guizhou plateau is, therefore, different from the towers of the Guilin karst. This may be due to lithological differences between the Guizhou and Guilin limestones – there being no equivalent in Guizhou of the sparry allochemical limestones of the Rongxian formation; the limestones in Guizhou are generally thinner-bedded and often intercalated with clays. Lateral undercutting at the water table will locally steepen the cone hills, but the strong vertical joints and massive bedding planes are missing – so slopes slump much more readily than in the Guilin karst. Steep-sided isolated tower-like hills occur in the Falang limestones (Triassic) of the Longgong area.

Photo 14. Cone karst (plateau and gorge) on the Guizhou plateau, west Guizhou

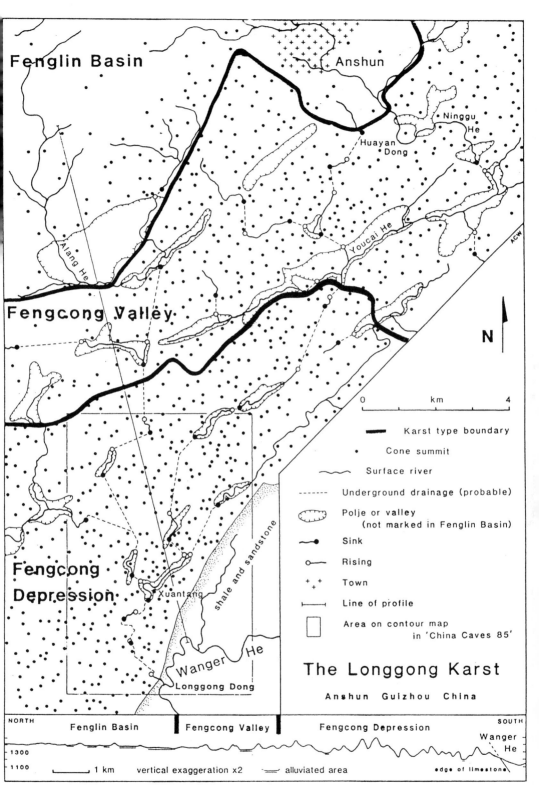

Fig. 28. The Longgong karst. (Eavis and Waltham 1986)

Handel-Mazetti visited the cone karst of Guizhou (1927) and referred to the cones as "inverted spinning tops". H. Lehmann also referred to Handel-Mazetti as depicting "a cone karst of wonderful regularity near Louping in Guizhou", (Lehmann 1936a,b). Lehmann further quotes Handel-Mazetti – "What is it that produces forms of weathering such as these cones...? The region is too remote from my experience for me to be able to attempt to answer that question, but my feeling is that the tropical-type rainfall conditions may have been a contributory factor." (Handel-Mazetti 1927, p. 285; Lehmann 1936b transl by M. Fargher).

The cones occur in definite lines or belts on the Guizhou plateau, separating basins or wide valleys. The lines of the cones are probably the relics of the original interstream divides and may also reflect lithological divisions. Between the cones are cols, the old karstified gullies or valleys; the cols are often U-shaped, which has given rise to the idea by some Chinese geologists that glaciers have modified their shape. The karstified plains and valleys on the plateau surface are relics of the original relief formed at a very early stage. The lines of the cones also indicate that the original relief was relatively steep and formed probably after a rapid uplift of the land.

The cones and the enclosed depressions on the plateau were able to develop over a long period. The closed depressions near the watershed zones can be very large, while those nearer the deeply cut valleys are much smaller. Lateritic crusts occur in many of the superficial sections on the plateau surface, however, no study of these has yet been done. Their occurrence suggests that the old plateau surfaces might have developed under warmer and even more seasonal conditions than those which exist today and when the plateau of Guizhou was at a much lower level. Waltham suggested that there was repeated burial and re-excavation of much of the plateau karst (Waltham in Smart et al. 1986).

5.1 The Cone Karst of Shuicheng

An important study of the cone karst on the plateau at Shuicheng, W. Guizhou, was made by Xiong Kangning (1985, 1992). In this area, which is over 2000 m a.s.l., the Sancha he (tributary to the Yangtse) and the Beipan river (tributary to the Zhujiang) are deeply incised into the plateau surface, up to 500 m; they and their tributaries are actively eroding headward. The climate in Shuicheng is cooler than the more easterly parts of Guizhou – average annual temperature is about 12.3 °C and the average annual precipitation is 1230 mm/year. The cone karst is developed in Lower Carboniferous to Middle Triassic shallow sea facies limestone – mainly bioclastic textures. They are very pure and generally not very dolomitic. Shuicheng lies within the NW-SE fold zone of W Guizhou with close folds and near-vertical (compression-shear) faults. Dips are high – usually more than 50°, and often vertical and

The Cone Karsts of Shuicheng

reversed. The area is cut by several active Quaternary faults tending NW-SE, with their accompanying shear faults NE-SW. There are frequent fault-block movements, giving rise to karst fault basins and tablelands.

The fenglin landscape in the Shuicheng area consists mainly of isolated cone karst hills resting on a flat rock surface (Photo 15). They can be connected to each other by doline type depressions, valleys or basins. They vary in height from about 15 to over 100 m. Cones of this type are usually regarded as having been formed under tropical conditions in the Neogene (Smart et al. 1986). Though Shuicheng today does not meet the temperature conditions for the tropics, it has a high precipitation, about 1250 mm (Xiong Kangning 1992). The Quaternary climatic changes, in whatever ways they affected Guizhou, would also affect the cones – forming perhaps some of the debris forms which occur at the foot of the cones. In the Shuicheng area, there are planated karst surfaces at 1850–1900 m and a surface of cone summits at 2050–2150 m which tilts to the southeast. The highest surface is at 2400–2600 m.

Xiong Kangning (1985, 1992) chose four sample districts in the Shuicheng basin representing the fenglin types of fengcong depression, fenglin depression and fenglin plain. He measured the parameters of 745 cone hills, all of which were more than 15 m high. The cones exhibited a highly symmetrical form with a slope range of only 45°–47° and a mean of 45.9°. Only when there were intercalated clays or shales did the mean slopes decrease. Moreover, the slopes of each individual cone are also very symmetrical; there were no cases

Photo 15. Symmetrical cone on the Guizhou plateau (conical fenglin)

in which the slope angle on one side of the cone exceeded that on the other side by more than 2.99°, usually it was only about 1.025°. The diameter/height ratios of the cones were between 2 and 3.99. The planimetric slopes of the cones have a tendency towards circularity. The orientation of the cones is determined by the regional lineaments and fractures – not minor joints. The general slope of the land, particularly if it corresponds with the tectonic strike, influences the long axes of the cones which migrate towards the slope directions. The symmetry of the cones is expressed in terms of plan lengths of the two arms of both the major and minor axes which intersect at the cone summit. There are six possible degrees of symmetry (A-F, see Fig. 29). The cone symmetry is expressed by the symmetry product = $(L1/L2)(S_1S_2)$. When PKC is 1, the cone is symmetrical. In the Shuicheng region, most cones have values of less than 1.25. The symmetries of types E and F are created by the cones being part of a dissected escarpment. The depressions between the cones form cellular networks like polygonal karst. All the possible types of symmetry are given in Fig. 29.

From these results, Xiong Kangning attempts to find a sequential development of the fenglin cone karst. Accordingly, in principle, the series must be "in the direction of larger depressions, lower relief and smaller residual hills". There is no reason to suppose that the different fenglin landscapes come from the same geomorphological sequence – "we should consider them representing different stages of different sequences" (1985; Fig. 30).

The following conclusions come from the analyses according to Xiong Kangning –

1. The cone angles are stable and do not change with the dimensions of the cones, altitude, or lithology or attitude (dip) of the rocks. This shows that the slopes of the fenglin are a time-independent, open system landscape in steady state.
2. Cone length/cone width ratios are significant in the four areas; the ratios on steep dips are generally greater than those on gentle. The greatest ratio is

Symmetry class	Symmetrical			Asymmetrical			Total	Symmetry Product Pkc	
Diagrammatic plan	A 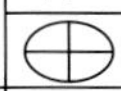	B 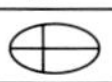	C 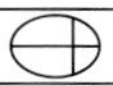	D 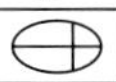	E 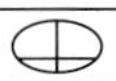	F 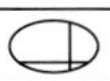	number of cones counted	Mean	Standard Deviation
Sample Areas	%	%	%	%	%	%			
Shuanglongjing	11	55	4	17	7	6	165	1.22	±0.23
Yuhuangdong	29	29	5	22	7	8	122	1.38	±0.32
Zhongba	8	41	2	23	3	23	321	1.75	±0.76
Fajing	15	38	6	23	6	12	137	1.49	±0.51

Fig. 29. Types of symmetry in the cone karst of the four sample areas in the Shuicheng area, western Guizhou. (Xiong Kangning 1992)

in the direction of slope of the land; the greatest ratios are also along the major fractures. From the fengcong depression, through fenglin depression to fenglin plain, the planimetric shape of the cones has a tendency towards circularity.
3. Constant cone diameter/height ratios determine the rate of the cone volumes as they diminish and become more homogeneous. In Shuicheng, the main difference in the resistance of the rocks is their resistance to corrosion – which changes not only with carbonate composition, but with fracture density and scale in the rocks.
4. Great cone length/width ratios imply that structural control is important; as the length/width ratio is reduced, the cone tends towards circularity and structural control is weaker, and karstification stronger. These ratios decrease from the fengcong depression through fenglin depression to fenglin plain, indicating a greater effect of karstification through the series.
5. Cone density increases throughout the sequence.

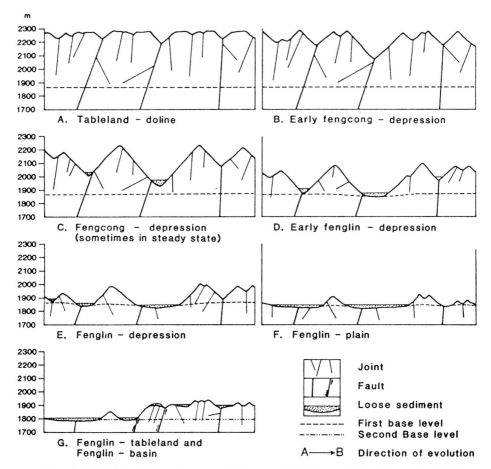

Fig. 30. Model of fenglin landscape evolution near Shuicheng. (Xiong Kangning 1992)

6. The cone distribution changes throughout the sequence from uniform to clustered and the areas of the polygonal depressions become greater. Those depressions with favourable conditions will capture those with less good conditions, which results in the cones becoming more clustered and random. This conclusion contradicts the work of Day (1979) and, in some ways, the writer's views of the breakup of the dry valleys. It is possible that this evolution comes at a very late stage of karstification – or in a critical stage of tectonics versus karst corrosion.

He further draws conclusions about the evolution of the cone (fenglin) landscape. The cone slopes are always stable in fenglin development. There is a strong linear, positive association of cone height versus diameter for each sample. The cone slopes lower synchronously and retreat parallel to one another. The distribution of the cones, fenglin, has a tendency from close to sparse. The distances between the nearest neighbour increase as the cones become more circular in shape; isolated cones are more circular than closely packed ones. A corollary of this is that connected cones tend to become isolated cones. The cone saddles (or cols) become lower as the distances between the cones increase. The fengcong depression landscape proceeds towards fenglin depression. In the early stage, there is greater accord of summit level, higher absolute elevation of the cones, and a greater cone length/width ratio along the structure lines. As the cone planimetric slope becomes circular, the summits lower slowly, but the depression floors are lowered more quickly and the cone relative height progressively increases. As this happens, the slopes of the cones remain constant, while the cone density decreases and the depression areas increase. The autocatalytic effect of the depression development makes them deepen with increasing speed; the number of cones will decrease, the distances between them increase, and the cone saddles (cols) lower and the cone summits (belonging to the captured depressions) will achieve a different summit level. However, this sequence is not unlimited as new swallow holes will develop in the depressions and result in the subdivision of the older, large depressions. Xiong Kangning notes that such competition might result in equilibrium, whereby a polygonal network of depressions is formed and there is a uniform distribution of cones. This situation will lead to a steady state landscape and the cone length/width ratio will tend towards 1:10.

The further development of the fengcong depression will depend upon the thickness of the vadose zone, which is controlled by regional tectonism and karst base level. The thickness of the vadose zone above the karst base level is similar to the available relief. Where tectonics are stable (static) and the karst base level is shallow, deepening of the depression is inhibited and lateral corrosion becomes important. Cone/summit and saddles are lowered by corrosion. When depression bottoms and cone saddles reach karst base level, fengcong depressions break down and turn into fenglin depressions. However, with regional uplift, and with a thick vadose zone, a steady state can develop. This is because there is some internal self-adjustment amongst the variables –

so that the fengong depression stays in a steady state with a polygonal network. Karst landscapes in this state can maintain the same form and same composition for some time, even though corrosion continues. For example, depression area and sides, cone volumes, slopes, relative heights, density and distances between cones all remain constant, while the elevation of the cone summits, saddles and depressions are progressively lowered at the same speed.

The fenglin-depression landscape has a tendency towards the fenglin plain. Xiong Kangning reports that further development of the fenglin-depression landscape will depend upon the parallel retreat of cone slopes. This will leave a residual low karst pediment at the cone base and further increases the depression area. This pediplain is like a karst planation surface. At this stage, the depression bottoms are lowered at a very slow rate, while the cone summits are lowered more rapidly. The cone volumes decrease, the depression areas increase and the distances between the cones also increase. The cone distribution also becomes more sparse, so that the many isolated small cones stand up in a fenglin plain. He concludes that the fenglin plain in the areas sampled were controlled by an earlier base level of the Sancha he, which was on the level of the present depression bottom before its rejuvenation and down-cutting. As a result of the present strong down-cutting many dolines developed in the depression floor (Smart et al. 1986).

The degree of geological control to these landscapes would *theoretically* become less and less and the karstification more and more, as the sequence fengcong depression through fenglin plain develops. After the sequence has reached the mature stage of fengcong depression with a polygonal network, the structural control of the landscape decreases to a minor role; the parallel retreat of the cones slopes in dynamic equilibrium will not lead to any changes in distances between the cones on the plain or to their migrations on the plain. However, in practice, geological factors do intervene; for instance, big cones may break down along major fractures, and the intervening space may or may not develop into depressions. As a result also of the parallel retreat of the slopes, small cones are destroyed more quickly than large ones and cones in the softer and more fractured rocks are destroyed more quickly than those in resistant rock. Xiong Kangning embodies these ideas in a model of landscape development (Xiong Kangning 1992; Fig. 30).

There are two main problems regarding these ideas. First, in the early stages of the sequence, the development of a gently sloped doline plain into steep-sided cones is still unexplained. According to Xiong Kangning's model, the dolines finally merge, the ridges are worn down to a fine pyramidal form, with symmetrical slopes from 45°–47°. This process has yet to be observed, and does not solve the problem which has puzzled all karst geomorphologists – how do the cones originate in the first place? In the writer's opinion, this is where fluvial erosion and gullying are very important in providing the original steep slopes and in giving the spatial dimensions of the landscape (Smart 1988; Fig. 30). Once the original valleys have broken up, the polygonal network of depressions, as described by Williams (1972) and elaborated by Xiong

Kangning, become self-limiting; each depression covers the space available as completely as possible. If the karst base level is low and the vadose zone thick, there are no sequential changes with time and the polygonal depression network and the cones are in a steady, time-independent state in dynamic equilibrium. At this stage also, provided there is sufficient relief above the regional base level, the evolution of these landforms on the surface of the plateau is decoupled or detached from the effects of base level and the underground drainage.

Secondly, the lack of geological control in the middle and later stages discussed by Xiong Kangning would seem to be most unlikely in the field of geomorphology where differences in lithology, corrodibility, and structure faults are so important. As in the Guilin area, once the details of the texture and lithology of the rocks are known, the importance of the geological controls become evident and are fundamental to any interpretation of the relief. Until such detailed studies are done of the limestones in the Shuicheng area, it is difficult to accept the absence of geological controls. After all, apart from the effects of the hydraulic gradients, dependent on tectonics, karstification is completely dependent upon the nature of the limestones, not only in their corrosivity, but also in their mechanical strengths. It seems that the uniformity of the cones (slope angles etc.) in the Shuicheng area must depend upon uniformity and homogeneity of the limestones which are being attacked by erosion and corrosion and that this must be one reason why the cone karst of Shuicheng is so spectacular. If the cones orginate in the first place as divides between fluvially drained gullies, the original spacing of the relief could be conditioned by initial irregularities in the terrain, as well as geological factors (Fig. 31). As time goes on, geological factors are, therefore, likely to become *more* important in the spacing of the drainage net. Tan Ming suggested a two-tier arrangement of relief; the upper tier is in dynamic equilibrium, the lower tier is in an evolving form. Both forms co-exist in the same landscape (Tan Ming 1992a,b; Photo 14).

5.2 The Canyons of Guizhou

The effects of the tectonic uplifts on the Guizhou rivers have been profound. Canyons cut into the plateau and are from 500–600 m up to 1000 m deep. Gullies intersect the slopes, and hanging karst springs and caves are frequent. In the valley of the Maotiao river, the outlet of an underground river hangs nearly 200 m above the valley, giving rise to an impressive waterfall (Fig. 27). Siphon springs are frequently found. The activity of karst water in the canyon area is intense and karst conduits are relatively well developed. Compared with the relatively slow evolution of the karst on the plateau, headward erosion and corrosion are rapid. The canyon area increasingly expands, until it replaces the plateau landscape. Headward erosion and corrosion are thus a basic process, and have three compenents: (1) headward development of the surface drainage system; (2) headward development of the underground drainage system; (3) headward development of the karst landscape.

1. Headward Erosion. The headward erosion of the surface drainage is seen in the series of waterfalls and rapids (possibly knick points) in the longtitudinal profile of the streams (Fig. 32). One of the best known is the Huangguoshu fall (Orange waterfall) on the Dabang river, studied by Zhang and Mo (1982) and mentioned by Xu Xiake in 1683 (Photo 16). The Huangguoshu fall is over 80 m

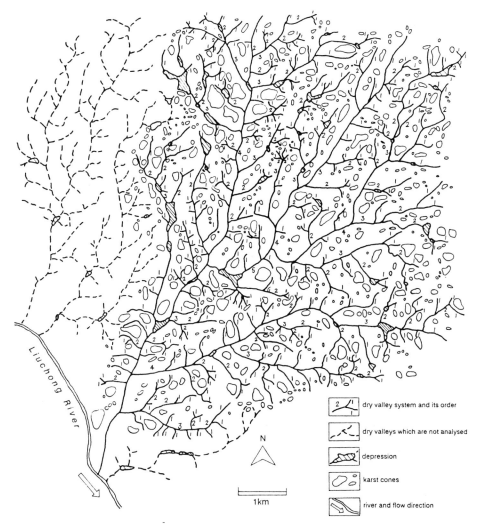

Fig. 31. Association of karst cones and depressions with a well-developed dry valley system. (Tan Ming 1992)

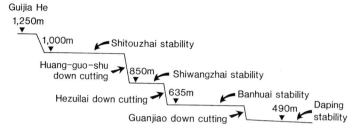

Fig. 32. Geomorphological stages of the Dabang river. (After Zhang Yingjun and Mo Zhongda 1982)

Photo 16. Huangguoshu waterfall, Guizhou

high and like many of the waterfalls is accompanied by a thick deposit of tufa (Fig. 33). Tufa is formed partly by the degassing of CO_2 as the water falls – but this process is probably also assisted by calcareous algae and mosses. The Huanggoshu separates the two types of karst geomorphology – the Shanpen

The Canyons of Guizhou

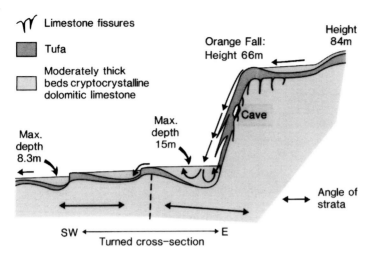

Fig. 33. The Huangguoshu fall (Orange waterfall), Dabang river, Guizhou

(mountain basin stage) with an area of planed and subdued cones and the Xiagu canyon stage of deep dissection. The canyon below the Huanggoshu is 400 m deep. In general, the rivers disappear underground on approaching the knick points (steep parts) and reappear when the gradient of the riverbed becomes more gentle. In the case of the Huanggoshu, the main waterfall is but one of a series of falls in the Dabang river and the river disappears about 5 km below the main fall – disappearing into the rocks and giving rise to a natural bridge, the Tian Sheng Qiao, which is about 0.5–1 km long. Natural bridges are a frequent phenomena at or near the knick points (see also the natural bridges discussed by Smart et al. (1986) in the Shuicheng karst. Great blocks and chaotic collapses indicate that under conditions of high water, the Dabang river does not entirely disappear into the conduits underground and that a surface stream flows. A kilometre or so below the disappearance of the water, where the gradient flattens out, the river emerges in a fine resurgence (Photo 17) 31 m deep. This is a further indication of the formation of shallow phreatic caves – just below the water table. Huangguoshu is in Lower and Middle Triassic limestones and dolomites. It is regarded as a regional knick point by Zhang Yingjun and Mo Zhong da (1982).

In the eastern part of Guizhou in the neighbourhood of the town of Zhenyuan, the plateau is dissected by the Wuyang river. The limestones are Ordovician and Cambrian and are much intercalated with shales and other clastic rocks. The canyon of the Wuyang, though only 100–300 m deep, is very narrow, like a cleft in the limestone rocks. Part of the canyon has been dammed to form a narrow lake-like reservoir. Above this rise pinnacles of weathered dolomitic limestones. Tributary valleys enter the canyon by steep, hanging valleys; tufa deposits are found where waterfalls occur. The tributaries occasionally shift their junctions with the main stream, leaving old

Photo 17. Large spring in the Dabang river

deposits of tufa suspended on the canyon walls. This can be seen in the valley wall opposite to the town of Zhen yuan.

The Maotiao river, a tributary of the Wujiang in the central Guizhou, shows similar features. Its total length is 160 km, its gradient is 1.6% in its upper reaches but 6.6% in the lower reaches. The upper section is wide and open, the lower section deeply incised in a canyon (Fig. 27). The Wujiang river itself is graded directly to the Changjiang (Yangtse), but its tributaries, like the Maotiao, have waterfalls in their upper and middle reaches (He Caihua 1982; Zhang and Mo 1982). The Maotiao river being near to Guiyang city has four hydroelectric stations built along its course – one on the plateau surface, the others in the gorge sections. Along a single river course, steep falls may occur at several levels, the gradients between the steep parts being relatively gentle. The rivers, therefore, flowing off the plateau may disappear underground and reappear several times. Some observers believe that the actual breaks in the river profiles represent the actual intermittent uplift of the earth's crust – though in reality they may coincide with structural geological boundaries.

2. *Long Profiles of the Underground Rivers.* The long profiles of the underground rivers also show such steep breaks. The longitudinal profiles of the underground streams steepen towards the resurgence, and at the Longgong cave in the Anshun area, the break point, the Longmen waterfall, is at the resurgence (Fig. 34). Underground river capture is also common as a result of the steepening of the gradients. As a result of the uplifts also, former old phreatic caves systems are exposed in multiple layers – indicating a long erosional history.

3. *Headward Development of the Karst Landscape.* The karst landscape in the canyons becomes progressively younger with increasing distance upstream of the drainage system. However, on the plateau, the sequence is from older to younger – as the plateau surface is rejuvenated (Fig. 35). The morphological landscape is characterized by the geomorphological zoning already shown in Dushan. From the water-dividing regions on the plateau in the direction of the canyon are found: peak forest valley or basin (Fenglin basin); peak cluster (Fengcong); shallow depression; peak cluster deep depression (Fengcong). Along the canyon sides are found strips of fengcong depression – the strips varying from a few kilometres to over 10 km. In the depressions, the water table may be very deep – up to 100–200 m; in the depressions, vertical shafts are better developed than horizontal caves, in contrast to the fenglin-plain areas, where the tiers of horizontal caves predominate. There are numerous dolines and sink holes in the depressions – the linear distribution of the sink holes reflects the flow of underground water. After heavy rain the depressions may be inundated, because unlike the plateau, the transmissivity of the sink holes is not sufficiently high to allow rapid drainage of the water. There is intense erosion by the water and underground collapse is frequent.

The zonation of the landforms is not sharply defined and is more a broad concept dependent upon the differential tilted blocks of uplift, and geological controls, more than an evolutionary cycle. There was no unified plateau surface over this great area, but presumably a series of uplifted tilted blocks, each with its own karst sequence. If the landscape sequence of the Guizhou plateau is linked to that of the Guangxi basin, a multilayer karst landscape sequence is obtained: gently undulating karst hills, peak forest depression, peak cluster depressions, peak forest valley, isolated peak plains, as in Guilin (Zhang Zhigan 1980a).

The zonation is particularly well shown in the Anshun-Longong karst area (Fig. 28). The structural grain is NE-SW and the fenglin basin karst of Anshun occupies the broad interfluve area between the Sancha he (to Yangtse) and the Wanger He (to Zhujiang). The suite of the landforms from the fenglin basin to the fengcong depression area by the Wanger He is essentially continuous – all being initiated at the same time. The contrasts are due to the uplift along the Wanger He. Rejuvenation has not yet reached the Anshun plateau and yet the interfluve landscape there is regarded as more "evolved". The Anshun relief consists of rolling hills, 200 m high with isolated cones and stretches of quite

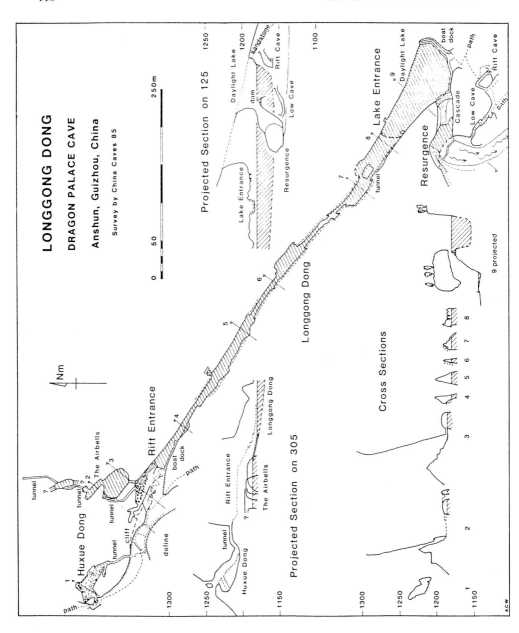

The Canyons of Guizhou

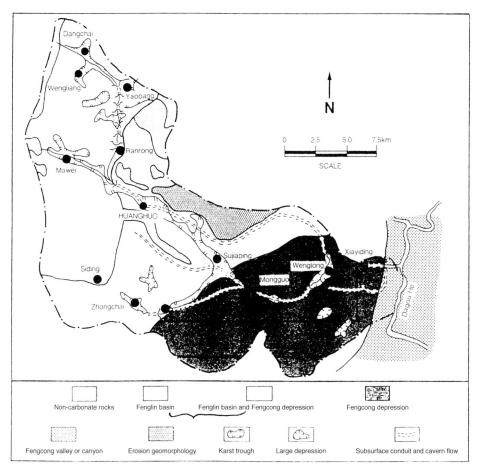

Fig. 35. Map of the main karst geomorphological zones in the catchment area of the Huanghe underground drainage system, Dushan area. (Song Linhua 1986)

flat areas; the caves have a low gradient through the isolated cones and ridges. The water table is shallow, the aquifer is highly transmissive, consisting of fissure flow plus occasional isolated phreatic conduits. In the fengcong depression karst the regional water table has dropped very low, there is an absence of surface streams and the longest caves are here. North of Anshun is a fengcong depression area which the China caves group surveyed; an example is the Long Tan shaft, 275 m deep, one of many shafts breaching the floors of the fengcong depressions. The caves show falling base levels, abandoned passages, and cave levels of phreatic origin, with present-day vadose modification (He Caihua 1982; Waltham 1984).

◄──

Fig. 34. Longgong Dong (Dragon Palace cave), Guizhou. (Eavis and Waltham 1986)

The caves of the Longgong karst, surveyed by the China caves group (1986; Fig. 34) include Longgong Dong (Dragon Palace cave), now a tourist cave. The resurgence of the Longgong stream is over a waterfall (now used as a small hydroelectric station) above the Wanger He. The cave is in the Triassic Falang limestones and the resurgence is caused by the limestones being brought up against the shales and sandstones on which the Wanger He flows. The Longgong resurgence has a discharge of 480 l/s in the dry season and 5500 l/s in the wet season. The cave entrance is via a perched lake, the site of an old phreatic "lift"; the show cave is a single phreatic rift, now half-flooded and it can be traversed for 600 m by boat. The rift cave ends in a collapsed entrance overlooking a deep fengcong depression and a network of dry valleys. Steps and paths lead on to the Yema Dong, where the tourist path ends at present. In the Yema Dong, there is a 150 m-long gour slope with a deep sump pool 70 m below. The drainage of this area of the Longgong karst is "in and out" (Fig. 28) with the underground sections being interrupted by surface streams in closed poljes (Photo 18).

Photo 18. Polje, Longgong karst, Guizhou

All this work on Guizhou stresses the distribution of the landforms in regular replacement – a unique regularity which is the geomorphological expression of the hydrodynamics brought about by the neotectonic movements. Each of these groups evolve in its own time – so that each fengcong depression or fenglin basin is of a different age and not synchronous.

5.3 The Caves of Guizhou

As expected, caves are important components of the Guizhou karst. An illustrated book on the caves of Guizhou and also a memoir on the caves have been published (Zhang Yingjun 1985). Many caves await exploration, though quite a number are already known and also commercialized, as at Longgong Dong. As a result of the uplifts, caves appear in tiers of two to three and sometimes as many as five layers, as for instance, the Monk Cave on the left bank of the Yuliang river. In caves that are now abandoned, stalagmitic and stalactitic deposition is important as in the Guiyang Underground Park (White Dragon cave) in the neighbourhood of Guiyang city and in Daji (Zhijin) cave in the Yulong mountains. The Daji (Zhijin) cave is in the Maokou limestones and is 3.98 km long (Yang Mingde 1982b; Zhang Yingjun and Mozhongda 1982).

There are two kinds of cave systems in Guizhou: (1) tunnel caves with an allogenic catchment area, this group includes the caves in Guizhou; (2) there are also caves in folded geological structures with an autogenic catchment, which are more characteristic of Hubei and Sichuan. The tunnel caves of Guizhou "are the most numerous and voluminous of this planet. Many subterranean canyons surpass 100 m in height (120–150 m at Gebihe). The big room at Gebihe (Ziyan, Guizhou), called the chamber of the Miaos, measures $700 \times 200 \times 70$ m and is the second largest known in the world (Barbary et al. 1991; Fig. 36). Zhang Shouyue, Philippe Audra and Richard Maire also comment, "The genesis of the large underground features is directly related to the impressive discharge of the allochthonous rivers passing through the tunnel caves. The giant pits are due to the collapse of large underground chambers or galleries. In the remnant stumps of tunnel-cave passages, levels are spread across a vertical distance of several hundred metres and have developed in the course of the strong uplifts during Neogene and Quaternary" (1991; Fig. 37 and Photo 19).

The carbonate deposits of many of the Guizhou caves have been dated by U/Th disequilibrium series by Zhao Shusen et al. (1988). They are all less than 300 000 B.P. Between 120 000–70 000 B.P. there was much deposition, while around 150 000–170 000 B.P. there was little deposition. There are no dates yet available for the tufa deposits that are found associated with the waterfalls on the rivers in this area. In the Zhijin cave, many of the stalagmites resemble those in Aven Armand (S France) and Postojna cave (Slovenia). Zhijin cave in the Sancha he area, west of Anshun is developed in Lower Triassic limestones, which are quite thinly bedded. As a result, many collapse features occur in the cave itself and also in the valleys of the Sancha he and its tributaries, "natural bridges" are a feature of the Zhijin cave area. Huaxi park, S of Guiyang city, is another conservation district, with some fine tufa waterfalls.

5.4 The Slope Zone of Maolan

The slope zone between the Guizhou plateau and the Guangxi lowland lies on the SW part of the Yangtze platform. Much of the folding is generally NE-SW. An unusual area of forested peak cluster (fengcong) karst exists in Maolan Nature Preserve, near the town of Libo (Guizhou province). The nature preserve consists of Carboniferous and Permian limestones and dolomites oriented NNE-SSW with dips from 5° to 10°. The Maolan karst varies from 1078 m a.s.l. in the NW to 430 m in the SE towards the Guangxi lowland. The area is unusual in that it is more than 90% forested – the forest being made up of evergreen and deciduous broad-leaved mixed trees (Zhou 1986). Despite the forest cover, rocky limestone outcrops are abundant and on the slopes of the peak cluster hills 70–80% of the bed rock is often exposed. The relief between the top of the peaks and the base of the depressions is from 150–300 m. Rainfall is about 1750 mm/year falling mainly from April to October, and the mean air temperature is about 15 °C, 5.2 °C in January and 23.5 °C in July. Because of the relatively dense forest, the mean air humidity is high – over 83% on the peaks and 94% in the depressions.

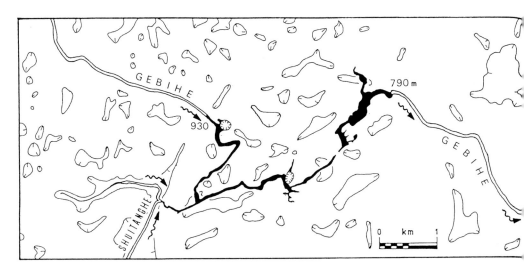

Fig. 36. The underground course of the Gebihe and Gehibe cave. Closed depressions are also shown. (Barbary et al. 1991)

The Slope Zone of Maolan

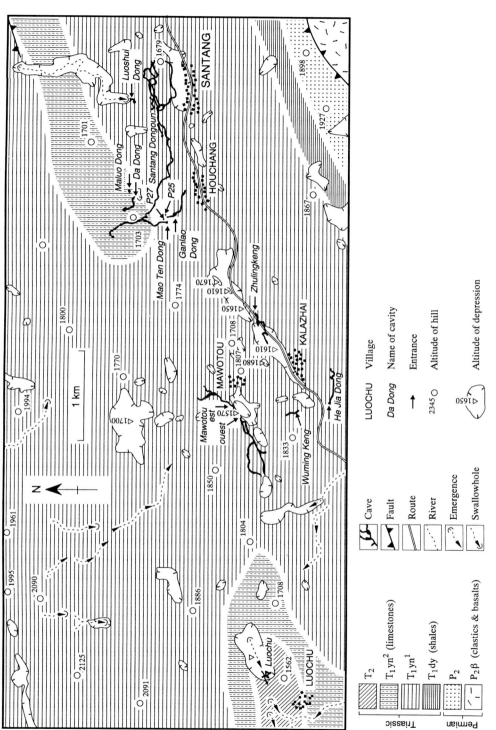

Fig. 37. Geological context of the Santang syncline and localization of caves (Zhijin-Guizhou). (Barbary et al. 1991)

The vegetation shows a vertical change in the deep dolines or depressions (Liu Jiming et al. 1991). Soil CO_2 values are high and superficial corrosion of the limestones is important. Because of the large quantity of CO_2 from plant remains and humus, epikarstic fissures are well developed; epikarstic water is an important component of the hydrology of the karst. This can be seen in the Wangpaishan peak cluster depression. The epikarstic water is a regulatory part of the hydrological system and this compares markedly with areas where the vegetation and soil have been removed. In the soil eroded areas which are very typical of the rest of the Guizhou plateau and of Suicheng, drought and waterlogging are more prevalent. As a result of studies in the Maolan forest area, Chinese scientists are now well aware of the regulation of the hydrology brought by the forest to karst development and exploitation. In the forested Maolan district, there is a double-layer hydrological structure – the epikarst zone and the lower aeration zone of conduit flow; this structure mitigates the problems of drought and waterlogging in the karst.

Recent neotectonic uplift has affected the Maolan area. This has produced a deep drop in the underground water table and helped to give rise to the

Photo 19. Cave entrance, Guizhou

dissection and development of the cone-like peaks and the intervening depressions. The peak clusters are close together. The diameters of the depressions are usually 200–400 m but can be as much as 1000 m. As a result of the tectonic uplift, many of the rivers are now underground; springs emerge at the faulted junctions of the limestones with the sandstones and shales of the lower Permian, as at Wolong. The surface sections of the rivers drop rapidly and tufa formation is frequent at the waterfalls. The rate of tufa deposition is at present very high – some tens of centimetres being formed in 20 years. Some of the older tufa has been dated by ^{14}C to 2699 years. Tufa deposition is believed to be caused by both degassing of the CO_2 in the water and also by bryophytes absorbing the CO_2 from the water. Despite the Maolan area being a National Nature Preserve, small hydroelectric power station developments are taking place along the Huanghou stream in the neighbourhood of the Wolong emergence. Each of the different karst environments in Guizhou have different environmental and engineering problems – these are discussed by Zhang, Yang and He (BCRA 1992) and in the series of essays, *A Study on the Karst Environment in Guizhou,* edited by Yang et al. (1988)

5.5 The Age and Origin of the Karst in Guizhou

Guizhou is a much larger area than the Guilin area and detailed studies are still scarce. It is, therefore, difficult to generalize about the overall evolution of the Guizhou karst. There is no doubt, however, about the effects of recent uplift. The dating of these uplifts is believed by Chinese geologists to be late Tertiary and also well into the Quaternary – some believe that up to 1000 m uplift occurred in Tibet and SW China in the mid- and late Quaternary. Not all the faulted and folded blocks of Guizhou have been equally affected by the neotectonics, but several active blocks like that in the Shuicheng area exist. The separate structural blocks probably reacted differently to the tectonic stresses. Each structure has a separate evolution both geologically and geomorphologically and with a separate sequence of development of the karst landforms. The geomorphologists in Guizhou believe that there are stages which are common to all the structural blocks as stated by He Caihua (1982) on the basis of work by Yang Huairen.

1. *Dalou mountain stage* from late Cretaceous to early Tertiary. During this period the fold mountains of the Mesozoic Yanshan stage were much eroded. Sediments and conglomerates of early Tertiary age accumulated in basins, some of which have now subsided. The highest peaks of Guizhou, as in W Shuicheng, are regarded as remnants of this early period.
2. *The mountain basin stage* from the Miocene to early Pleistocene during which time the main karst landforms now seen on the various plateaux surfaces were formed. This long period of time allowed the formation of large karst basins, fenglin basins, wide shallow valleys and karst cones. This period was believed to have had a tropical climate or a climate warmer than

now exists in Guizhou and the thick red crusts, that are found on the plateau and in the poljes, were formed at this stage. Large polje lakes, horizontal caves and shallow karst water were also characteristic. On the slope zone of S Guizhou, peak forest landforms were developed at this time.

3. *Wujiang stage* According to He Caihua, strong uplift of the crustal blocks began in the middle Pleistocene. As a result of the rejuvenation by the rivers, the land surface of the mountain basin stage is being destroyed. The uplift has been intermittent, so there are accompanying river terraces and caves. The Wujiang has four to five terraces and three to four layered caves; the terraces of the Wujiang river are 30–35, 60, 80–90 and 190 m above the present river. He Caihua believes these terraces represent four temporary pauses in the Quaternary uplift.

The karst of Guizhou shows the basic regularities of the development of relief on a rapidly rising land surface during a period of climatic change in the late Tertiary or Quaternary. We know very little of the climate in Guizhou during the Quaternary, but there is evidence of some dry phases, probably during the glacials (Shi Yafeng 1992). There is little evidence of frost or snow action in any part of Guizhou, so there is not much foundation for any periglacial or glacial landforms. The Sino-French group reports some periglacial deposits in Ziyan area (1991). As a result of a possibly cooler climate and the rapid Quaternary uplifts, fluvial erosion of the limestones is dominant in Guizhou. Fluvial erosion accounts for the initial development of the present-day cone karst and the strong outlines of the original valleys which are still seen today. Even today, fluvial erosion is dominant; it is estimated that the rate of lowering by fluvial erosion is 310–980 mm/1000 years, while the rate of karstic (solutional) lowering is only 33.8–77.6 mm/1000 years. Because fluvial denudation exceeds the rate of karst corrosion, the original karst network is being broken up. This is, therefore, a contrast with the state of the karst in the Guangxi basin (Song Linhua 1980; Yang Hankui et al. 1984).

Smart et al. (1986) in their consideration of the geomorphology in western Guizhou (particularly Shuicheng) combine the effects of downcutting of the rivers and the effects of uplifts on the development of the karst. This combination, in many ways, is the essence of the Guizhou relief, though the working out of the details has not yet been done. Furthermore, Guizhou illustrates several aspects of karst development. It shows that an original cone-like karst is unlikely to develop into a tower karst. It also indicates the ways in which the karstification of the valleys takes place, and the sequences in the development of karst depressions and the establishment of a polygonal hydrographical network. The development of polygonal type depressions and their particular

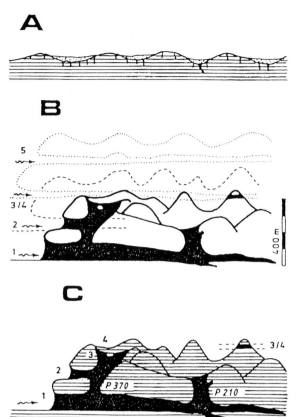

Fig. 38. Schematic geomorphological evolution of the Gebihe area during the Cenozoic in relation to the Himalayan uplift. **A** Eocene morphology; **B** Oligocene-Miocene to Pleistocene (cave levels *1-2-3-5*); **C** the Gebihe today. (Barbary et al. 1991)

hydrology will not usually take place unless there is a sufficient superficial alluvial or other thin cover of soft rocks upon the limestones. Polygonal karst depressions need the widespread retardation of the surface water for their optimum development. A schematic evolution of the tunnel caves of West and Southwest Guizhou is given in Barbary et al. (1991; Fig. 38).

6 The Karsts of Yunnan

6.1 Introduction

Yunnan is the most westerly province of S China and nearly 26% of the whole province of 97 000 km² is karstic. The easterly parts of Yunnan form the western sectors of the Yangtse platform with folds generally NE-SW and NNE-SSW. However, from Kunming, the capital of Yunnan, westwards, the structure lines are N-S or NNW-SSE (Figs. 3 and 4). Western Yunnan is the region where the eastern Himalayas and the mountains of Assam and Burma impinge on the South China paraplatform and where high ridges separate the deep valleys of the rivers which have risen in Tibet – the Mekong, Salween, and Irrawaddy and the Red river which flows to Hanoi. There are three main areas of karst in Yunnan: (1) in the west and south in the N-S ranges adjacent to the Burmese and Vietnam borders; (2) the graben or karst basins aligned N-S in the Kunming and Nanpan river areas; (3) the undulating plateau area in E Yunnan which is a continuation of the Guizhou plateau, but at a higher altitude where the famous "Stone forest" of Lunan is situated. The Indo-Sinian (or Neo-Cathaysian) folds affected the rocks in Yunnan as they did in Guizhou and Guangxi. However, the effects of the Cenozoic and later the Quaternary uplifts are much greater in Yunnan than in the more easterly provinces. The proximity of W Yunnan to SE Tibet is the most important aspect of the relief. Carbonate rocks of the Palaeozoic and the Triassic are predominant in E Yunnan; in W Yunnan, the carbonates are mainly Permian and Triassic, but in some areas, older metamorphic limestones (Ordovician) occur, as at Dali.

The name Yunnan means south of or below the cloud. Compared with the cloudy climates of Guizhou and Sichuan, the Yunnan climate with its sub-tropical air is much sunnier. Kunming, at an altitude of 2000 m, is regarded as having a climate of "eternal spring" with an average annual temperature of 15 °C, ranging from 20.5 °C in summer to 8.3 °C in winter. The rainfall, which occurs largely from May to October, is about 940 mm/year.

West and S Yunnan. This area is largely tropical, and along with Hainan Island, S Guangxi (Nanning), the most tropical in China, lat. 22–23°N. It includes the Xishuangbanna district. Here, there are two main karst areas one in Mengla county and the other along the Lancang river, the upper part of the

Introduction 121

Mekong river. Permian and mid-Carboniferous limestones are important in the Mengla district while along the Langcang river the limestones are mainly bio-clastic and dolomitic limestones of the Triassic. Both areas are intensively folded and faulted. Karstification has been active over a long period of time and continued during the Quaternary. Both the Mengla and the Lancang river areas have peak cluster karst with depressions, blind valleys and underground streams. The relief is steep and the gradient of the surface and underground rivers is also steep with multiple cascades. Several caves are recorded but probably many more are waiting to be discovered. Part of the Mengla area is in tropical rain forest and part of a Tropical Nature Preserve (Menglun).

The Xichou karst of SE Yunnan (lat. 23°05′) has been studied by Song Linhua and Liu Hong (1989). This area is in the transition zone from the Yunnan plateau in the north to the lower mountains and hills of N Vietnam. The average annual precipitation is 1260 mm with most of the rain falling in May–October. The average annual temperature is 16 °C. The limestones extend in age from the Devonian to the Upper Triassic, and the total carbonate sediments are over 4000 m thick. This area is transitional from the folds of S Guangxi and S Guizhou to the N-S structures of W Yunnan and the geological structures are arc-like. Under a warm climate in the Tertiary period, purple and yellow clays were deposited on the limestones.

The karst resembles that in the neighbouring parts of Guangxi and Guizhou. It has experienced considerable uplift during the Quaternary, and some of the rivers are quite deeply cut into 300–400 m canyons. Fossil or relict karst landforms are preserved on the plateaux above the river valleys. The arrangement of the depressions and the cone hills of the fengcong show how they are related to the fractures and faults and less to the folding – though the cone hills form lines that are parallel to the strike of the limestones (Fig. 39). Because of the neotectonics, three levels of erosion surfaces can be recognized and three terraces along the main river the Chouyang, tributary to the Red river (which flows into Vietnam).

The folding is more intense further west in Yunnan closer to the Burmese border and the longitudinal mountain ranges higher. The N-S ranges in the west can reach 4000–5000 m in the Hengduan Shan and the Nu Shan. These ranges also contain karst rocks, presumably affected by frost weathering at such high altitudes. There are few studies from this remote area as yet. From accounts by geomorphologists who have worked in this part of West Yunnan, the high alpine karst seems to resemble the high altitude karst in the Minshan in N Sichuan (Chap. 8) with frost-weathered pinnacles at high levels and remarkable tufa and travertine deposits at lower levels in the high valleys.

The Faulted Basins of Central Yunnan. The faulted basins or grabens of Yunnan are related to the compression and collision of three major tectonic plates – the Tibetan and the Yangtse massifs and the Indian plate. The largest basin is the one in which Kunming is situated and which is over 1100 km^2 in area (Yuan Daoxian 1991). The Kunming basin is controlled by a series of

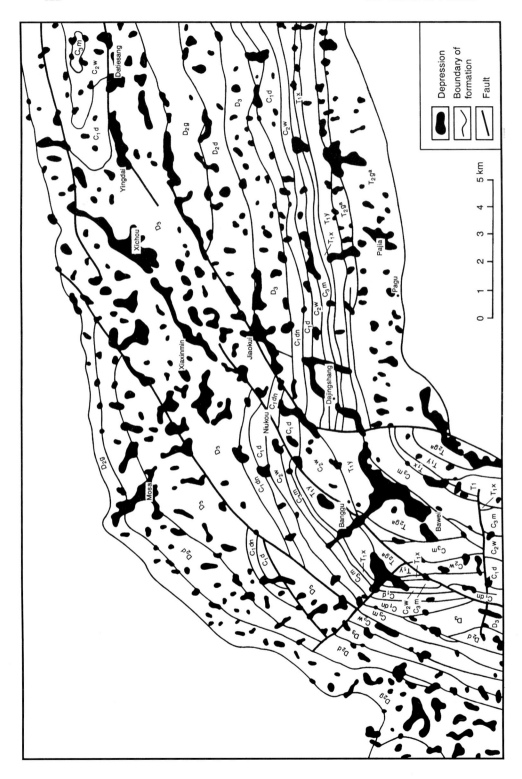

Introduction

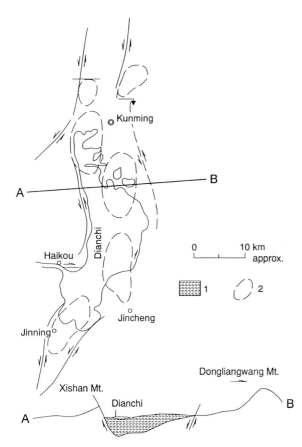

Fig. 40. Structure of the Kunming basin; *1* basin filling; *2* secondary minor basins. (Yuan Daoxian et al. 1991)

Tertiary and Quaternary normal faults, the downthrow sides of the normal faults have subsided to form a graben (Fig. 40). Palaeozoic and Sinian limestones have been downfaulted and the Kunming basin contains over 1000 m of overlying loose Tertiary and Quaternary strata. Many of the basins contain lakes, the most famous being the Dianchi lake in the Kunming basin, which is 40 km long N-S, and 8 km E-W with an average depth of 5.5 m. On the west side of the Dianchi lake is the Xishan fault which is an active fault scarp 500–600 m high (Fig. 40). The lakes are alkaline (pH over 9) and artesian water often issues from the underlying karstified limestones (Chan et al. 1987). In the Kunming basin, hot springs with temperatures up to 40 °C occur in the Sinian limestones. Beneath the lakes the downfaulted limestones are highly karstified and have a well-developed hydrographic network. This high degree of karstification contrasts with the relatively young karstification of the elevated limestones in the uplifted fault blocks and their bounding fault scarps. For instance, the west wall of the Dianchi lake is a very straight cliff with no

◄──

Fig. 39. The relationships between depressions and structure in S Yunnan; *T* Triassic; *C* Carboniferous; *D* Devonian. (After Song Linhua and Liu Hong 1989)

indentations and only very shallow caves of a few metres depth. The cliff shows little signs of solution, despite the moderately high rainfall (940 mm/year). Occasionally, the Dianchi fault scarp has broken into fragments of caves of the former hydrographical network. The shallow caves on the cliff side may have been formed during the recent period of active faulting and may be related to recent water tables or even to lake levels. Some of the small caves on the west wall of the Dianchi lake are now tourist attractions and also religious sites. The Dianchi lake is a permanent lake, as also are Yanzhong and Tonghai.

There are about 28 graben basins similar to the Kunming basin in Yunnan. Another well-known one is the Dali basin which lies to the west of the others and contains over 2000 m of Tertiary and Quaternary sediments. The lake at Dali, the Erh Hai, is one of the largest in the graben area. Some of the limestones at Dali are metamorphosed and of Ordovician age: they have been quarried for decorative purposes for many years and are world-famous. The Gejiu-Mengzi area south of Kunming in S Yunnan is also characterized by faulted basins, and lies between the two deeply cut canyons of the Honghe and the Nanpan Jiang. In the Kaiyuan basin the karst aquifer is 1000–2000 m thick; the Cenozoic sediments are also considerable in Kaiyuan.

The lakes, or *haizi*, in these grabens can be classified as permanent or temporary, and resemble the poljes of the Dinaric karst. Erh Hai is a permanent lake, but others dry out in the dry winter and spring seasons. The Banggu Maizi is covered with 3 m of water in the rainy season, but is dry for almost 3–4 months. Works to regulate the flooding and drying out of water in the *haizi* have not yet been very successful (Song and Liu 1989). The morphology and the hydrology of the graben basin karst in Yunnan resembles parts of the Dinaric karst – in which the geomorphology and hydrology are strongly controlled by faulting as in the Livno and Gacko poljes (Yuan 1991). However, in the Dinaric karst the faulting is generally older than that in S and E Yunnan and the hydrological networks in both the poljes and between them seem better established.

The Karst Plateau in E Yunnan. The karst plateau of western Guizhou is continued into East Yunnan, but at a higher altitude of 1800–2000 m. The Yunnan plateau is rolling and undulating. Limestones from the Devonian to the Permian are covered by Eocene (and possibly Miocene) lake muds and clays and a deep lateritic soil cover. Where these covering clays are being stripped and eroded, the underlying limestones become exposed. The structures in the limestones are open synclines and anticlines with dips from 3°–17°; fissures are closely spaced. Where the limestones are thick-bedded and strongly jointed, rock columns up to 30 m high can be formed. Groups of these columns are known as stone forests or *shiling*. The most famous stone forests are in the Lunan and Lingzhi areas, SE of Kunming. The stone pillars are developed in the lower Permian Maokou (354 m thick) and the Qixia (100 m thick) limestones. Both of these limestones are very uniform in composition and thickly bedded. The Maokou is a platform sparitic and bioclastic limestone with beds, according to Song Linhua, up to 30 m thick (Song Linhua 1986b). It is slightly dolomitized in places. The Qixia is a massive reef limestone and also dolomitic. The relief of the plateau in the area of the stone forests is about 100–200 m and the highest pillars occur in the shallow karstic depressions (or uvalas) which dissect the plateau.

The stone pillars or pinnacles are close set, a few metres or less apart, and usually over 5 m high. The height of each pillar is about three times its diameter. Where they are less than 5 m high, they are often called stone teeth, or stone shoots, particularly if they taper upwards. As the overlying Tertiary red mudstones and siltstones are removed, widened fissures in the limestones are exposed and the rock pillars isolated. The shallow dish-like depressions are most favourable for the development of the pillars and where they can be 40 m high. On the gentle hill slopes, the pillars may be 10–15 m and stone teeth characterize the higher intervening areas. The outlines of the pillars are related to the directions of the jointing, the main joints at the Lunan stone forest being N20°W, N50°W and N50°E. The N20°W joints are pre-Cenozoic and are filled with calcareous tufa, but the other joints are Cenozoic in age and are open vertical intersecting fissures, which cut the subhorizontal Maokou limestones. Stone pillars never occur in highly dipping limestones. In the depressions, the stone pillars are often separated from each other – but on the slopes and hilltops can be more closely set. On the summits on the planation surface and on the slopes where the overlying clays have been eroded, the pillars can be seen emerging from the clays. These clays can be shown to be in situ and undisturbed. The Shiling stone forests are crypto-karsts – so designated by Nicod and Maire (Barbary et al. 1991; Nicod and Fabre 1982; Photo 20).

The Lunan stone forest is situated at an altitude of 1650–1875 m; precipitation is 936 mm/year and the average annual temperature is 16.3 °C. Between June to October 70–80% of the rainfall falls. The Lunan stone forest occurs E of the Jiu-Xian-Shiyakou fault part of the Xiaojiang faulted belt (Fig. 41). East of the fault zone the ancient erosion surface is preserved, but to the west the surface is actively dissected by deep gorges as in other parts of the Guizhou–Yunnan plateau. The Ba river, S of Lunan begins to dissect the Lunan plateau in response to the cutting down of Nanpanjiang; the Nanpan flows in a gorge over 300 m deep. Near the confluence of the Ba river and the Nanpan there is a 90-m-high waterfall at Dadieshui. The top of this fall forms the base level for the Lunan plateau drainage, which has so far been unaffected by the rejuvenation of the Nanpan. The local water table in the Lunan area is represented by the Sword Peak pond, Lotus pond and Stone Forest lake, but surface and underground water drain to the Bajiang – to the S and SW. The underground water is in epiphreatic channels as in the subsurface stream which flows below the Lotus pond. Because of the variation in the rainfall during the year, the water table in the lakes like Sword Peak and Lotus pond may rise over 10 m during the wet season (Geng Hong et al. 1988; Photo 21).

The pinnacles in Lunan reach a maximum height of 30–35 m with many at about 20 m. The sides are vertical or nearly so. The tallest in this area occur in the centre along the drainage lines, while towards the ridges the clay cover

Photo 20. Karst pinnacles, Lunan stone forest, Yunnan

is greater and the pinnacles smaller. In other areas of stone forest, tallest pinnacles occur on the watersheds where the clays are removed and the limestones remain covered in the shallow valleys. The major joints between the pinnacles may be as much as 10–20 m wide, or only about 1 m wide. The majority of the fissures are from 1.0–1.5 m wide. Examination of the pillars, or pinnacles, suggests that several main groups of processes are involved in their weathering.

The *upper parts* of the pinnacles – the top 1–5 m – are dissected by different types of rillen-karren, and solution flutes (Photo 20). At least two sizes of runnels are formed. First, there are small, what might be called "standard" rillen-karren, similar to the rillen-karren described in many other environments. They vary from 1–2 cm wide, on projecting rock pinnacles they are several centimetres long; they are wider in the troughs (3–4 cm) than on the projecting pinnacles. Secondly, there are larger and smoother flutes or furrows up to 35 cm wide and up to 50 cm deep; these are from 1 m to several metres long. These longer flutes are characteristic of the Lunan pillars, and resemble solution flutes in humid and seasonally wet tropical areas such as Mulu in Sarawak, Borneo, or the Fitzroy region in Australia (Jennings and Sweeting 1963). At one locality in the Major Stone Forest, there is historical evidence to suggest that the larger solution furrows (i.e. 25 cm and more deep) have taken about 1000 years to form.

The upper parts of the pinnacles are also etched by scalloping (or cockling); scallops are shallow pits on both vertical and gently sloping surfaces

Introduction

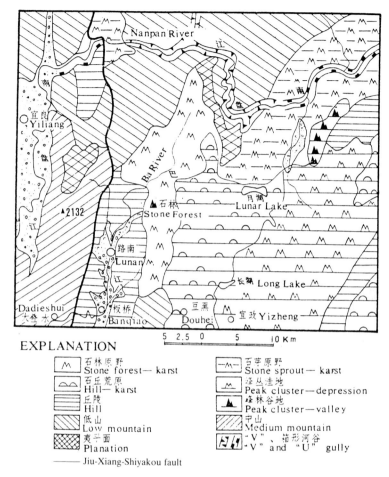

Fig. 41. Landscape map of Lunan. (IAH Conference Guidebook, Geng Hong et al. 1988)

caused by the action of flowing, eddying water; scallops are usually asymmetrical, the steeper side facing in the direction of flow – as in scallops in cave environments. As in caves, the scallops on the pinnacles are of various sizes, from 1 cm wide and 1 cm deep to over 30 cm wide and 15–20 cm deep, reflecting the speed and volume of the water and its hydraulics passing over the limestones. Small solution basins or dishes (*kamenitzas*) are common on horizontal surfaces and can vary from about 7 cm deep to larger features over 1 m long and up to 0.5–1 m deep. Many such solution basins contain algae and soil particles and are actively deepening today. There is thus plenty of evidence for active corrosion of the pinnacles today; groundwater contains up to 177 m/l $CaCo_3$ and 30 m/l $MgCO_3$, and the values are higher in the solution dishes. With enriched CO_2 the development of the weathering features on both the

Photo 21. Water table Sword Peok pond, Lunan stone forest

upper and lower parts of the pinnacles is affected by the presence of chert or dolomite in the limestones. Both chert and dolomite inhibit the formation of regular rillen-karren solution furrows and scalloping by interrupting the regular flow of water. Nodules of chert often have the effect of concentrating rainwater and causing the development of ring-like solution hollows around the nodules – a case of accelerated corrosion. The dolomitic limestones are also less soluble than the pure limestones, and weather more rapidly by physical processes (Chen Zhiping et al. 1986).

The lower parts of the pinnacles are more affected by vegetational and biological weathering, caused by the roots, stems, and leaves of tree creepers and other plants. As a result, the regularly fluted and furrowed upper parts of the pinnacles become replaced by irregularly pitted limestone and the rock is quite knarled in appearance. Some of the weathering differences between the various parts of the pinnacles may also be due to the strong lithological variations between the thick limestone beds. Such lithological differences account for the so-called mushroom rocks in the SW part of the stone forest,

Introduction

where variations occur in the incidence of fractures between the upper and lower beds. Other stone mushrooms occur when dolomite forms the lower beds; the dolomite, though less soluble than the pure limestone beds, is more susceptible to physical weathering and weathers more quickly than the upper beds. This phenomenon is well developed in the Lingzhi stone forest (Chen Zhiping et al. 1986).

Each pinnacle or pinnacle cluster is separated by narrow joint controlled corridors (or *bogaz*, see Cvijic 1924) normally 1–2 m wide and 20–30 m high. Solution along the major bedding planes takes place and limestone blocks become loosened. The blocks collapse and break down and the corridors widen in places, particularly at the intersection of major joints; steep-walled doline-like closed depressions are thus formed. Although the sides of the pinnacles are more or less vertical, a feature of their walled sides is the occurrence of smoothed semicircular recesses 2–3 m in diameter and as much as 20–30 m high; these recesses resemble the semicircular potholes or half-dome pits formed by the corrosion and erosion of concentrated falls or films of water, and which occur in limestone areas with and under loose cover rocks. Corrosion pits of this type are found in the Mammoth cave limestones beneath the Big Clifty Sandstone in Kentucky and in the Carboniferous limestones beneath the Pennant Grit in S Wales. Their existence was first discussed by Pohl (1955). It is possible that the semicircular features in the walls of the stone forest pinnacles had a similar origin beneath the cover of Tertiary sands and shales as those pits in the limestones formed beneath the cover rocks in Kentucky and S Wales. A discussion of such features is given in Song Linhua (1986b; Fig. 42).

As already mentioned with regard to the stone teeth and the pinnacles, the red Eocene mudstones and sandstones can be seen in situ, with gentle dips, undisturbed between the protruding limestones. A particularly good site to see this is near the Bimu pond in the SW of the Major Stone Forest; however, undisturbed and bedded cover rocks can be found throughout the area. Stone

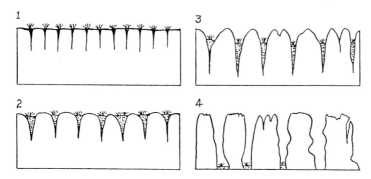

Fig. 42. Evolutionary stages of the stone forest according to Song Linhua (1986b). *1* Original fissured limestone; *2,3* subsoil erosion to enlarge the fissure and form the stone teeth; *4* stone forest development

teeth and pillars emerge from the Eocene beds and their associated lateritic deposits, indicating their formation by solution beneath the cover rocks. The cover rocks are sufficiently permeable and thin to allow the passage of unsaturated and corrosive water through them. It is, therefore, clear that the pinnacles and stone teeth have been largely formed in the first place by subsurface solution and that they are an example of subjacent or interstratal karst (*unterirdische* karst) first described by Penck (1924), or *cryptokarst* (Nicod and Fabre 1982; Nicod 1992a; Photo 22).

As the limestones are exposed and the overlying clays removed, subsoil corrosion is replaced by the corrosion forms of rainwater and surface vegetation. The limestones corroded by subsoil water are usually fairly smooth, according to Yan Qintong (1982) with a "subtle rough surface", often with residual dolomite powder. The limestones corroded by direct rainwater are characterized by sharp flutes and karren, rain pits, and scalloped solution forms; the vegetation clinging to the pinnacles produces more roughened hollows and irregular furrows. At the contact of the soil and clay-covered part of the pinnacles with the atmospherically weathered sections, there are troughs and irregularly perforated rocks. The subsoil corrosion is probably enhanced by mixing corrosion, which enlarges the fissures between the pinnacles and gives rise to long funnel forms (Song Linhua 1986b).

Yan Qintong and others have estimated the modern solution rates in the stone forests (1982). In general, the speed of corrosion of the pinnacles in the

Photo 22. "Stone teeth" (crypto-karst) in Lunan stone forest pinnacles of Permian Mao-Kou limestones becoming exposed by the removal of Eocene clays

Introduction

depressions (uvala areas) is ten times the speed of corrosion on the hill summits; this is due to the richness in biogenic CO_2 of the soil water in the depressions. Calculations have been made of the total amounts of calcium and magnesium in water flowing away from the pinnacles. The rate of lowering by solution on pinnacles on the hilltops is approximately 29 mm/1000 years, and on those in shallow depressions it is 29.2 mm/1000 years. These calculations are supported by observations on a stone pillar broken by the Tangchi earthquake in 1833, which now has karren from 3–7 mm deep, cut into a fresh surface; from this, Yan Qintong calculated that the solution rate on the pillar is from 2–3 mm/1000 years. He also estimated that the rate of elevation of the stone pinnacles in the centre of the basins is about 280 mm/1000 years but is only 26 mm/1000 years at the margins of the basins. If the present rates have continued over a long period of time, then the elevation rate or period of formation of the stone pinnacles (up to 30–35 m high) in the centre of the basins would be from 100 000–175 000 years. However, other factors, in addition to corrosion, have been important in the pinnacle formation and it is also unlikely that the corrosion rates would remain the same over such a long period of time. Hence, these figures may not be a true indication of the pinnacle age.

Most of the water in the stone forest depressions does not drain over the surface to the rivers, but sinks underground. The rates of calcium and magnesium dissolution under these circumstances are high and the waters are often supersaturated (over 300 mg/l). When the water issues into the rivers as springs, tufa precipitation occurs. Tufa precipitation on the second terrace of the Ba river for instance, is from 5–15 m thick; the base of this tufa has been dated by Zhao Shusen using the U/Th method to be about 40 000 B.P. (Zhang Shouyue 1984). A tufa at the base of a deposit by the natural bridge (Tianshenqiao) at the entrance to the tourist part of the stone forest has a date of 175 000 B.P. The tufas contain faunas which are of late Pleistocene age. Tufa is also associated with Dadieshui waterfall, formed by the Pleistocene uplift of the plateau. The tufa is late Pleistocene in age. Chinese geomorphologists have recently begun to study the incidence of tufa formation and its relationships to waterfalls; the role of organisms (mosses and bacteria) are recognized in tufa formation, but studies also indicate that degassing and loss of CO_2 is a contributory factor (Yan Qington 1982).

The geological evidence suggests that the Lunan and Lingzhi pinnacles have their origin in a covered karst. New pinnacles and stone teeth emerge as the covering clays are stripped away. Zhang Shouyue (1984) shows how stone teeth or shoots develop into stone pillars – though there are some workers who believe that the stone teeth develop from degraded stone pillars. If the soil cover is stable, the bases of the pinnacles will be eroded and eventually the pinnacles will collapse. Substantial alluviation will eventually cover some of the pinnacles, leaving many isolated on the plateau surface (Photo 23).

The location of the Lunan stone forest was determined by the movements along the Juixian fault zone; these movements controlled the deposition of the Eocene rocks and allowed their gentle overlap on to the Permian limestones.

Photo 23. Later stages in the development of the Lunan stone forest – degraded Crypto-karst

Over 500 m of red mudstone was deposited in the Lunan region. Yan Qintong and others recognize three main stages in the development of the Lunan stone forest (1982; Fig. 43).

These are:

1. *The Lumeiyi stage.* This stage lasted from the Mesozoic to the late Eocene. For most of this time from the Mesozoic until the early Eocene, the area was subject to long-term denudation and planation. In the late Eocene, the Lumeiyi and Xiaotun red mudstones were deposited as lake muds upon the limestones. As the lake dried out, a network of streams formed on the clays and groundwater conditions allowed the limestone pinnacles to develop under the clay cover. The pinnacles formed at this time have approximately the same diameter and the spaces between the pinnacles are very narrow, possibly indicating that the stone forest area at this time was experiencing relatively dry conditions. The parts of the stone forest which were formed at this time can be seen in the eastern part of Dabushao mountain and at Dou Lei village; these localities were situated on the gentle slope of the eastern shore of the Lunan lake and the pinnacles formed at that time have subsequently been dissected and weathered.
2. *The Dacun Village Stage.* This stage extended from the Miocene to the end of the Pliocene and the beginning of the Pleistocene. In general, it was a period of denudation; the pinnacles which had formed under the Eocene clay cover particularly in the preceding Lumeiyi stage gradually became

Introduction

Period and Stage		Mould	Stage description
Plateau Period (K–E_1)		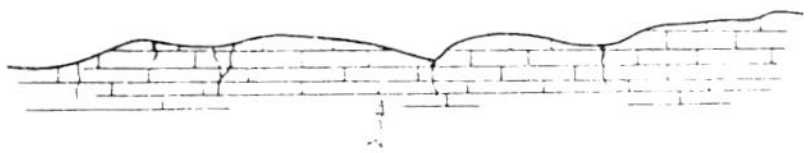	Stage of formation of old planation surface; Karst hills & depressions formed with red clay deposits.
Lake Basin Period (E_2)	I	Ancient Lake 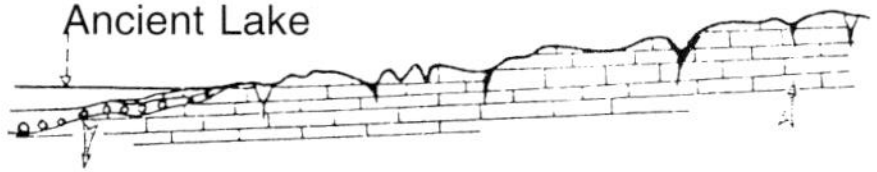	Stage of ancient human lake basin.
	II	Ancient Lake 	Expansion of lake basin stage. Karstification developed further in eastern part of area, climate dry and hot.
Broad Valley Period (E_3–Q) and Cutting Period of Ba River (Recent)	I	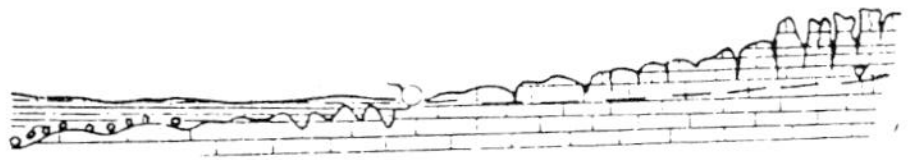	Development of Stone Forest I. Climate changing to moist tropical and sub-tropical
	II	Ba River 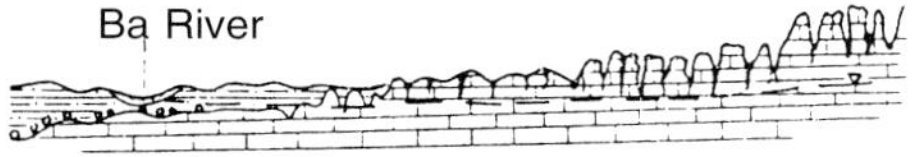	Sub-stage II of Stone Forest; trival valley of Ba river formed. Underground water-table falling.
	III	Ba River 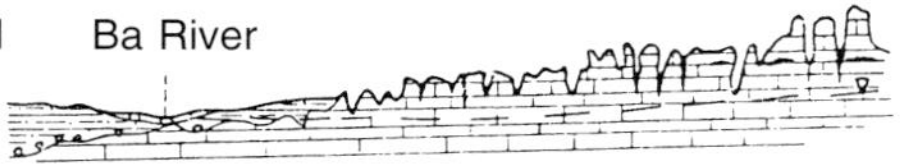	Stone forest I destroyed. Beginning of Stone Forest III.
	IV	Ba River 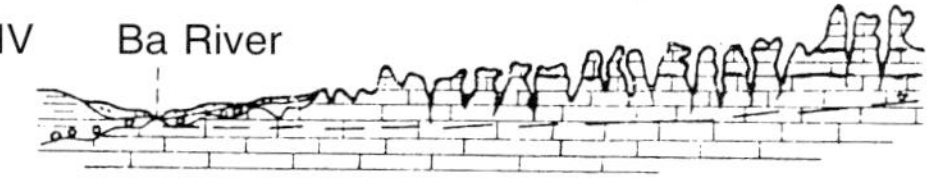	Stone forests II & III develop into rock pillars. Stage IV develops along valley.

Fig. 43. Stages in the formation of the Luan stone forest. K Cretaceous; E Eocene; Q Quaternary. (After Geng Hong et al. 1988)

uncovered and modified by atmospheric erosion. There was some general uplift of the whole area and the large shallow closed basins were intitiated but the underground water drained away slowly. A limited amount of headward erosion of the Nanpan river took place, but the hydrographic network was not well developed and the planation surface was not destroyed. The conditions did not favour the development of tall pillars in the stone forest, but pillars and pinnacles or stone teeth 3–5 m high formed under the cover rocks and soil.
3. *Lunan Tourist Area Stage.* This stage is believed to be middle and late Pleistocene. As a result of more intense uplift, the Nanpan river and its tributaries accentuated their downcutting and underground drainage became more vigorous. The large closed basins (uvalas) which were initiated in the preceding period became the sites of convergence of surface streams and concentrated seepage and provided the conditions for the development of the stone pillars. The upper parts of the pillars were subject to corrosion by atmospheric waters and lake waters; as a result of the further removal of the covering clays and soil, the relative heights of the pillars increased up to 35–40 m towards the centres of the closed basins.

Yan Qintong claims that the stone forests of the Lunan area can be arranged in three stages according to their relative heights above sea level. Thus, the pillars and pinnacles belonging to the Lumeiyi stage are 20–30 m higher in altitude above sea level than those of the Dacun stage, while the Dacun stage pillars are 5–8 m higher than those of the Lunan tourist area stage. Though Yan Qintong regards the Pleistocene period as being the most important in the development of the Lunan stone forest, his explanations involve the subjacent solution of the limestones beneath the Ecocene clay cover over a considerable period of time.

Other Chinese writers, notably Yu Jinbiao (1982), state that the main periods for the formation of the pinnacles of the stone forests were in the Eocene and early Oligocene. According to this hypothesis the pinnacle karst was more or less formed before the deposition of the Tertiary lake clays. However, it is unrealistic to assume that no change and no dissolution of the limestones took place during the time they were covered by the lake clays and soils. It is also important to stress that modern (post-Pleistocene) corrosion and erosion of the stone pillars are quite significant.

A map of stone forests similar to those at Lunan, occurring in other parts of South China, is given by Song Linhua (1986b; Fig. 44). The stone forests in Guizhou (at Puding and Meitan) and at Xingwen in Sichuan are also in Maokou Permian limestones, a bioclasic series with beds up to 5 m thick. The stone forest at Buermen park, Yongshun, Hunan, is developed in dark pink Ordovician limestone with beds 1–3 m thick. The Xiaopaiwu stone forest in Huayuan, Hunan, is in dolomitic limestones with beds about 2 m thick. In all these areas, the beds are generally gently dipping, never more than 15°–17° and usually much less. The development of the pillars and pinnacles is control-

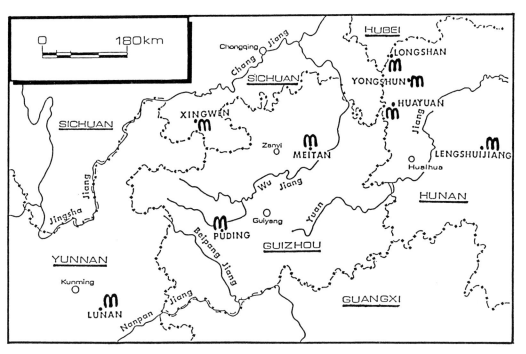

Fig. 44. Distribution of stone forests in S China. (Song Linhua 1986b)

led by the vertical joint and fissure network. The evolution of the pillars and pinnacles in these areas is discussed by Song Linhua, who illustrates the importance of their development under a soil and loose rock cover. He stresses the importance of unsaturated soil water and underground water in the pinnacle formation, and indicates that though rainwater solution is an important modifier of the pinnacles, it is not the main agent in their origin.

Unlike the fengcong cone karst areas of Guizhou and other parts of S China, the stone forests developed largely by corrosion beneath a relatively loose rock cover. The influence of fluvial denudation in the origin of the present karst forms is, therefore, much less. The pinnacles of the stone forests are just emerging from their cover rocks. Compared with the geologically old landforms of the Guilin tower karst, the stone forest landforms are much younger. Stone forest pinnacles occur where there is a loose rock or soil cover and where the bedding, lithology and jointing of the limestones are suitable for their formation, and where the weathering conditions are favourable and where the old plateau surface has been uplifted but not yet deeply cut into by the dissecting and rejuvenated rivers. Under these conditions, the gradual stripping of the overlying cover rocks to reveal the *unterirdische* or crypto-karst will take place. The importance of the origin and development of karst under a cover of newer rocks is being increasingly recognized (Nicod 1992b; Salomon and Astruc 1992).

6.2 Conclusions on the Karst of Yunnan

Since Yunnan's climate is broadly subtropical and tropical, the main differentiation in the karsts depends upon the tectonic situation. In West Yunnan, the karst is developed in north-south oriented faulted blocks and basins. However, East Yunnan is a continuation at a higher level of the Guizhou plateau and of the South China NE–SW fold zone. Compared with Guizhou, many of the limestone areas in East Yunnan are still covered with or only just emerging from the Eocene red beds, giving an *unterirdische* karst or crypto-karstic landscape. This landscape is of great importance as it gives as an indication of the early development of the much more evolved karsts of Guizhou and Guangxi. Work by Nicod and his colleagues in France have shown the significance of "les karsts sous couverture" or covered karsts (Nicod 1992).

7 Karst in Other Parts of South China

There are other areas of peak forest/peak cluster landscape in South China, which though not as famous as Guilin, have some fine karst. It will be seen from the map of karst types (Fig. 15) that the peak forest type stretches in a broad band from the Vietnam border, near Nanning in the SW, to near the Yangtse delta near Shanghai. Not all of this area was covered by limestones – because of magmatic activity associated with the Pacific plate or as in some areas the limestones have been removed by erosion and the underlying basement is exposed. As indicated in Chapter 1, the present state of preservation of the soluble rocks depends upon the zones of uplift and subsidence, which have affected the eastern part of China. In the areas of Guangdong, Fujian and Zhejiang, which belong to the second uplift zone, the soluble rocks have been to some extent removed by denudation, or were never deposited and their real distribution covers only a small part of the region – usually in down-faulted basins. However, in the subsidence zone to the W in Guangxi, Hunan, Jiangxi and to the S (Anhui, Henan, Hebei and parts of Shandong and Liaoning in the N) 40–60% of the soluble rocks has been preserved, though often they are buried under a cover of newer rocks (see location map of S China, Fig. 45).

In the NE, near the Yangtse in Hunan, Hubei, Jiangxi, Zhejiang and Jiangsu, Zhang Zhigan and Ren Mei claim that the characteristic peak-forest landscape (between lat. 27 and 30°N) has now degenerated into a modified hill-depression type and into a modified "temperate" variety of karst (Ren Mei et al. 1982). The limestones of this area are also often in basins covered by Cretaceous red bed sediments. One of the most important basins is the Heng Yang basin. The stripping of these overlying sediments now expose emerging limestone pinnacles, particularly on the low hills. The railway line from Hangyang and Changsha, northeastwards to Shanghai, passes through much of this emerging limestone country. Not enough of the limestones is yet exposed to give rise to distinctive karst country – but the influence of the underlying rock is everywhere present in the crypto-karst. Buried karst also occurs in West Guangdong. In some areas where the limestones are exposed, the crypto-karst has evolved to a stone forest, *shiling*, similar to that in Yunnan. A stone forest at Dahu in Yong'an country, Fujian, has been described by Bao Haosheng et al. (1991).

As indicated in Chapter 4, the justification for discussing the Guilin karst area separately and also the areas in Guizhou and Yunnan is mainly due to

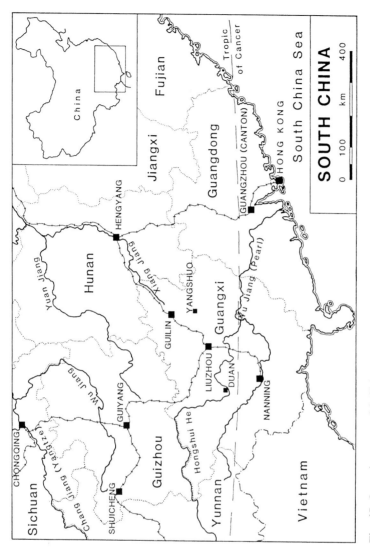

Fig. 45. Location map of S China

their distinctiveness and to the literature on those areas available to a Western reader. As the introduction to this chapter suggests, the karsts in South China (south of the Yangtse river) are regarded both by Zhang (1980a) and Yuan Daoxian et al. (1991) as tropical or subtropical. However, the limestone areas in the region of the Yangtse delta are quite different from those in southern Guangdong and South Guangxi. These differences may be partly climatic, but could also be due to lithological and structural differences in the limestones; they could also be the result of contrasts in the structural and tectonic contexts between southwestern China and the Yangtse-Nanjing area. These differences will be pointed out in this chapter.

7.1 Karst in Southern and Western Guangxi

In the South Guangxi and West Guangdong, where the limestones are preserved, peak forest similar to that in Guilin occurs. Some fine karst and caves are developed around the town of Luizhou where some of the Devonian limestones are particularly dolomitic (Fig. 45). Steep isolated peaks in the Liuzhou area are developed on stronger sparites and mingle with low conical hills on the dolomites. To the SW and W of Guilin much of the higher ground is made up of lower Palaeozoic non-carbonate rocks (sandstones and shales) which form the basement upon which the upper Palaeozoic carbonates (Devonian and Carboniferous) were deposited. The lower ground, which is followed by the railway line from Luizhou to Guiyang and the chief roads, is the Guangxi lowland and is largely developed on the upper Palaeozoic limestones. Large rivers like the Hongshui, Rongjiang, and Longjiang rise on the sandstone highlands and bring large quantities of allogenic water to the limestones. The karst is therefore largely peak forest (fenglin). The limestones have suffered fluvial planation and planed alluvial polje type relief is common, as at Yishan and at Jinchengjiang. Peak cluster relief (fengcong) occurs on the higher ground and where there was less influence of the allogenic water. This area forms the margin of the slope zone between the Yunnan–Guizhou plateau and the Guangxi lowland. All the rivers which flow S onto the Guangxi lowland are tributary to the Zhujiang (Pearl river); as a result of recent uplift, they are in deeply cut valleys and have conspicuous terraces from 20–40 m above the present river levels. They also have wide variations in discharge from the dry to the wet season (Liu Zaihua 1991).

Luizhou and the areas to the south of Guilin also are in the Guangxi basin, which lies in the second subsidence zone of East China. Both the Guangxi basin and the Nanning basin, which lies further to the south nearer the Vietnam border, differ from the Guilin basin. In these basins, the similar Carboniferous and Devonian limestones are relatively gently dipping, and the tectonics are less complex and more block-faulted. The faulting and folding of the limestones are probably Indo-Sinian (Triassic-Jurassic); the gentle basin formation of southern Guangxi is replaced in the Yunnan area to the west by block movements. The limestones of both the southern Guangxi and the Nanning basins were filled by Cretaceous and later terrigenous sediments, breccias, conglomerates and sandstones. There are also Permian cherty beds in the S Guangxi basin. The Cretaceous beds are often red beds resembling those found in the Guilin syncline.

The town of Laibin lies on the Hong Shuihe in the South Guangxi basin. In this area, though there are many Cretaceous deposits still remaining, the Carboniferous and Devonian limestones are well exposed and give rise to isolated fenglin and fengcong hills – particularly fenglin (peak forest) which rise to 300–400 m. The limestones in the Laibin area are over 5000 m thick. Where the Cretaceous red beds have been removed, the limestones have been planed by corrosion. One of the largest corrosional plains in S China occurs south of Laibin at Xiaopingyang and is over 1000 km^2 in area. The plain is at an altitude of 90–110 m a.s.l. with an average at 100 m. Occasional rolling hills of Cretaceous red beds can be seen resting on eroded limestone surfaces – very similar to the situation in the Guilin area – though in southern Guangxi the outcrops of the Cretaceous are much more extensive, and in the Laibin area they are over 200 m thick.

The Laibin area is drained by the Hong Shuihe and its tributaries. The Hong Shuihe rises in the karst and sandstone areas of Yunnan and Guizhou province to the north and has a very flashy karstic regime. In low water in the winter its level is only 40 m above the sea, but it can rise 40–60 m during the flood season. Its average annual discharge is about 2000 m^3/s but during the monsoon, the Hong Shuihe reaches over 20 000 m^3/s; during these periods, the river carries abundant red and brown sediment (its name, Hong means red water river) and resembles the Huanghe (Yellow river) in its sediment content. At its maximum the Hong Shuihe has a suspended load of 0.6 kg/m^3 on average, and 0.92 kg/m^3 at its maximum. Its suspended sediment per year is about 1/40 of that of the Huanghe.

The origin of the corrosional plains on the limestones is still debatable. The plains rise from the Hong Shuihe at a gradient of 1:1000. There are few allogenic streams in the S Guangxi basin and it may be that the corrosion of the limestones was brought about during the removal of the Cretaceous red beds. Water occurs in the limestones at relatively shallow depths in the corrosion plains – from 5–10 m deep. The surrounding fenglin and fengcong peaks contain dry caves, and there are many dry valleys. The limestone peaks give the impression of being very old static features in the landscape with little active erosion or corrosion taking place today, and of being even much more static than the hills and peaks of the Guilin area. The main aquifers in the Xiaopingyang–Laibin area are the Upper Carboniferous and the Devonian (Rongxian) beds. A series of gentle synclines in this Guangxi basin are water-rich. On the anticlines isolated fenglin occur, but as in Guilin, more fenglin occur in the synclines. In the Laibin area, the fenglin are fairly steep-sided as the beds are nearly horizontal, as in the centre of the Guilin syncline.

Further south, in the Nanning basin, the outcrops of the limestones become scarcer; the centre of the Nanning basin has Tertiary deposits overlying the Cretaceous beds. However, peak forest (fenglin) occurs, with ancient caves, some of which have yielded valuable remains. The most famous are the Yiling caves (Huang Wangpo 1977; Li Yanxian 1981).

West of central Guangxi, the amplitude of uplift in the later Tertiary and Quaternary was greater than in the Guilin area and is believed to be greater than 80–100 m. Western Guangxi is on the slope of the uplifted surface between the Yunnan–Guizhou plateau and the Guangxi plain – between the second and third step-like surfaces of China. As already indicated in the Guangxi area as a whole, the rate of uplift of the earth's crust is equal to or slightly less than the rate of karst denudation. However, to the W in the Duan region, the rate of uplift is equal or slightly greater than the denudation. The limestones in the Duan region are from the Middle Devonian through the Permian. The land surface remains well above the dry season groundwater table. Drainage appears in the valleys in the wet season only and, in peak cluster depressions, it appears when the highest flood level fails to reach the surface. In this area is the celebrated Disu river, which drains underground to the Qing Sui resurgence, a tributary to the Hong Shuihe river. This can reach a discharge of 400 m^3/s in the wet season.

The Disu underground river network is a good example of an underground hydrographic network in a peak forest area. It is on the boundary of peak cluster/peak forest area. It is the longest identified underground stream in China (241 km long) (Fig. 46). Because of the great variation in precipita-

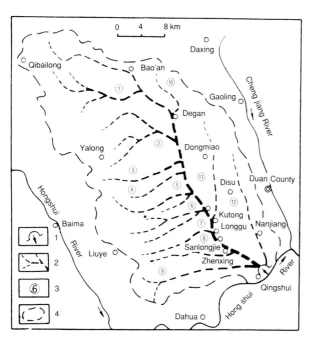

Fig. 46. The Disu underground river system. *1* Outlet of the subterranean stream; *2* subterranean stream; *3* number of the subterranean stream tributary; *4* the boundaries of the catchment area of the system. (After Chen Wenjun 1988)

tion from wet to dry season, the seasonal variations in the flow of the spring and underground rivers are over 100 times in the wet season than that in the dry season. In the Disu there are records of fast dye traces of 4 km/day and low hardness values of about 70 m/l. The China caves group (1986) remark that a well-developed conduit system extends behind the Qing Sui spring which may extend for over 30 km. Most of the underground cave system is flooded and there has been extensive cave diving to explore it. In this area, the denudation mainly deepens the valleys and the time has been insufficient for the development of the caves (Fig. 47), water levels in the Disu system can rise 80 m in 1 day (Chen Wenjun 1988).

If the rate of uplift exceeds the rate of karst denudation and the inherited hydrographic network is destroyed, erosion valleys will appear and the karst of mountain–valley–plateau type will be formed. This is a characteristic of Guizhou and of the slope zone between Guizhou and the Guangxi basin. The South Dushan anticlinal area occupies a transitional position between the elevated relief of the Guizhou plateau and the peak forest of the Guangxi.

In the late Tertiary and Quaternary periods, Dushan was elevated by about 500–1000 m, particularly in the northern part. As a result of this uplift, the north to south draining rivers (to the Zhujiang) now flow in canyons up to 200 m deep. Song Linhua (1986a) reported that the normal evolution of karst relief in an area like Dushan should be (from the drainage divides to the canyons) a doline-funnel depression type, fengcong depression and fenglin basin and plain in the lowest parts, using the zonation of Chen Shupeng (1965). However, Song Linhua shows in the Dushan area that the fenglin depression and fenglin basin relief are on the drainage divide. Song Linhua maintains that the reversal of the zones of the relief is due to the uplift and rejuvenation which steepened the hydraulic gradient and rejuvenated the karst in the lower reaches of the river basins in the neighbourhood of the canyon area. Old levels

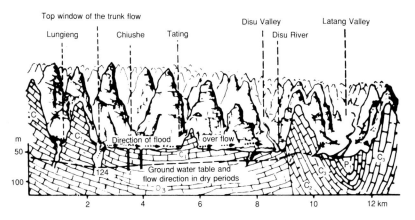

Fig. 47. The relationship between the trunk flow of the underground streams and the Disu surface river. (Hydrological team of Guangxi Geological Bureau 1976)

of phreatic caves are left as the water table falls. The relief in Dushan has, therefore, the stages shown in Fig. 48, doline (funnel) depression, fengcong depression, fenglin depression, to fengcong depression and fengcong canyon. Each of these different relief types is accompanied by a different pattern of groundwater flow. The zonation pattern given by Song Linhua is particularly well developed in the Huanghe catchment of South Dushan. Dushan lies along the railway line from Guangxi to Guiyang on the Guizhou plateau, where many of the types of relief discussed here can be seen (Gao et al. 1986, 1988; Zhu Xuewen 1990).

Zhu Dehao in Yuan Daoxian et al. (1991) compared peak cluster (fengcong) depressions along the slope zone bordering the Yunnan–Guizhou plateau, continuous with western Guangxi for about 500 km (see Table 14).

These peak cluster depressions are independent topographic units (often in domes of short axis anticlinal structures in Palaeozoic carbonates surrounded by Triassic sandstones) and illustrate the similarity of their morphometry and probably also evolution.

7.2 Karst in Hunan and Hubei and the Changjiang Gorges

Though Guizhou and Guangxi form the major areas of the South China karst, limestones are important in the adjoining provinces of Hunan and Hubei to

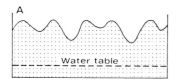
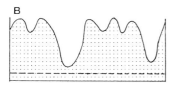
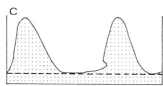
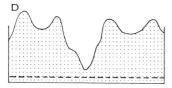

Fig. 48. Stages of karst geomorphology in South Dushan, Guizhou. **A** funnel and depression stage; **B** fencong-depression stage; **C** fenglin-basin stage; **D** spring-returned (fengcong-depression or canyon) stage. (After Song Linhua 1986a)

Table 14. Peak cluster depressions on the slope zone Yunnan-Guizhou plateau. (Yuan Daoxian et al. 1991, p. 63)

	Dafu (Guilin)	Sandu (Linjiang)	Longgang (Longzhou)
Elevation	656.6 m	662 m	580.8 m
Local base level	130 m	180 m	138 m
Total area of polygonel karst depressions examined	69 km^2	58.21 km^2	41.88 km^2
No. of depressions	182	115	269
R value	1.64	1.67	1.60
Av no. of sides	5.3	5.1	5.4
Average size of depression	0.38 km^2	0.51 km^2	0.153 km^2
Critical size	1.202 km^2	1.467 km^2	0.574 km^2

the E and N and in Sichuan and Yunnan to the NW, W and SW. Most of these areas are to the S of the Changjiang (Yangtse); the Changjiang cuts through the limestones in Sichuan and Hubei in the famous Yangtse gorges (San Xia). The Changjiang follows the general structural lines in the region near Chongqing, but between Wanxian and Yichang it cuts across the East Sichuan arc, as a result of the recent Tertiary and Quaternary uplifts. W Hubei is characterized by carbonate rocks of the Sinian, Cambrian, Ordovician and Triassic but particularly the Triassic and Cambrian. W Hunan has carbonate rocks of the same periods, but in addition there are Devonian and Carboniferous limestones. The limestones are intercalated with many non-carbonate beds, sandstones, shales, and basalts; the karst is of the *alternative* type compared with the *continuous* type of Guangxi where the limestone beds are much more continuous, and with few non-carbonate intercalactions. It will be seen from the geological map (Fig. 49 from *Gebihe*, 1991) that as a result of the folding and the sedimentary sequence that the carbonate rocks tend to occur more in longitudinal belts, separated by longitudinal belts of the non-carbonates. South of the Changjiang and also in Sichuan and Yunnan, where peak cluster fengcong occurs in such carbonate belts, Song Linhua has referred to them as *belted fengcong* or *band fengcong* (1986a).

The mountain and hilly areas, N and S of the Changjiang, the areas are karstic and the gorges resemble the gorge areas of Guizhou, but are deeper. In the Changjiang gorges the general trend of the relief is from W to E – from about 1800 m to less than 200 m. The karst is differentiated by the different rates of uplift along the Changjiang gorges and the Huangling dome. Karstic erosion forms are more important in the west, whereas the features in the east are transitional between the mountain and the plain. A section through the San Xia (Three Gorges) of the Changjiang is given by Lu Yaoru (1980) together with a table of comparison of the river terraces. The karst features are related to the base levels of the gradually uprising Changjing. Some of the

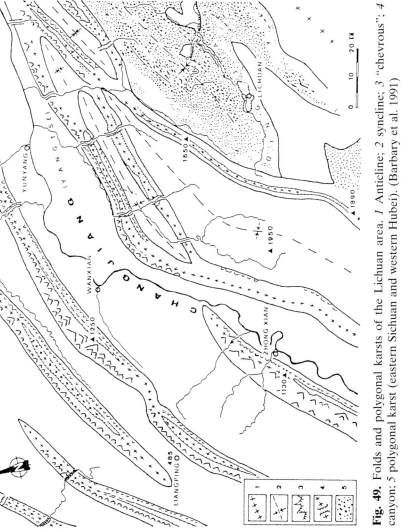

Fig. 49. Folds and polygonal karsts of the Lichuan area. *1* Anticline; *2* syncline; *3* "chevrous"; *4* canyon; *5* polygonal karst (eastern Sichuan and western Hubei). (Barbary et al. 1991)

karstic features of this area will be affected by the rise in water level caused by creation of the new reservoirs along the Changjiang; the effects of the development of the reservoirs and the new schemes are not properly known and give cause for some concern for the karst environment.

The geological structure of W Hubei is complicated because of the intensive folding and the carbonate rocks are separated into distinct patches. The karst occurs as isolated systems in either anticlines or synclines; this is illustrated by the Changyang anticline (over 100 km long) and the Yuntai Luang syncline in the area of the Qingjiang in western Hubei (Lu Yaoru 1987a; Yuan 1991). Each of these structural areas has developed its karst in isolation. Peak cluster shallow depressions occur in both W Hunan and W Hubei, but fluvial erosion is important and dominant in these areas. River caves are well developed in the Triassic limestones of the Qingjiang valley. One of the most famous is the Tenglong cave near Lichuan, Hubei, which has a subterranean river 10 km long (Dong Bingwei 1987). The cave has a high and a low level; the upper level is a former course of the Qingjiang, which now flows in the low level cave (Masschelein and Zhang 1990). Known are 37 km of cave passages which have been explored by Belgian speleologists. In both W Hubei and W Hunan, fluvial-karstic features (dry valleys, vadose caves) are abundant in addition to many kinds of closed karstic depressions. Flow in many of the underground passages and aquifers in Hunan and Hubei can be quite turbulent as indicated by the following figures (from Lu Yaoru, no date):

Yangtse gorges West Hubei	Coefficient of permeability km/day	Velocity of groundwater flow/day	Critical velocity of flow, VKP
	75	834–7488	<148

Linear underground rivers characterize the long synclinal or anticlinal passages in the limestones of West Hunan. Surveys of caves in this area have been done by the Franco-Chinese speleological group (Barbary et al. 1991).

South and East Sichuan are also composed of parallel folds and compressive strike faults. The anticlines are steep and narrow – about 3–5 km wide. The synclines tend to be broader and gentle, 10–30 km wide. Limestones of Triassic age occur in the anticlines, while Jurassic sandstones and shales occur in the synclines. Narrow karstic belts developed in the anticlines; these are similar to the Jura mountains of SE France. Breached anticlines form hilltop karst valleys and may be up to 110 km long. Strings of closed depressions formed along the hilltop valleys. As in the Jura mountains, gorges occur where the rivers cut across the anticlines. Many of the high depressions formed along the anticlines appear as karst lakes – at heights much above the surrounding sandstone and shale valleys; a well-known basin without water at 470 m and over 5 km long is known as the "pool in the sky" (Yuan Daoxian et al. 1991).

In SE Sichuan, the Maokou and Qixia formations of the lower Permian form an impressive karst area, between 540–820 m a.s.l. Mean annual precipitation is over 1200 mm and the average annual temperature is about 13 °C. The Maokou limestone in particular is massive and thickly bedded and strongly jointed. Solution and weathering along the planes of weakness in the limestones produced rocky columns up to 20 m high, similar to those, also in the Maokou, in the stone forest (Shiling) in E Yunnan (see Chap. 6). Removal of the overlying beds and soil erosion also gives rise to a crypto-karst and to what the Chinese call "stone teeth" or stone "shoots"; these are tapering stone pillars 2–4 m high and are separated by widened joints up to 3 m deep. The widened joints are kluftkarren which might be called grikes, as in N Britain; however, their formation is not associated with glaciation as are grikes in N Britain and the intervening stone teeth are quite different from the clints on limestone pavements. A feature of the lower Permian karst in SE Sichuan is the number of collapse pits – the largest one known as the big pit of Xingwen is over 650 m from E-W and over 490 m N-S and is 208 m deep. Steep cliffs form the upper half of the pit, while large collapse blocks cover the lower part; caves and an underground stream are connected with the collapse. Collapse features as large as this are relatively rare – two of the largest in the world occurring in the Dalmatian karst near Niksic in Croatia, the Blue and Red lakes; both of these have steeply cliffed upper sides with collapsed blocks below but they also contain relatively deep lakes (over 250 m deep). In the case of the Red lake, its base is believed to reach below sea level; it has been discussed by Roglic (1974). The Maokou and Qixia limestones in SE Sichuan also contain many fine caves, some of which have been explored (Brook 1993).

Both solutional denudation and mechanical denudation are important in E Sichuan, Hubei and Hunan. Lu Yaoru calls it the combination of corrosional/corrasion karst disintegration. He tried to quantify his ideas (Table 15) though the ideas and figures are to some extent hypothetical.

In central and southern Hunam, limestones of Devonian to Triassic age occur in moderately folded areas. Because of the fragmentation of the lime-

Table 15. Indices of disintegration

Area	West Hubei	NE Guangxi	W Guangxi	Sichuan	NW Hubei
Average thickness of corrosion in relation to average thickness of corrosion	0.0012	4.4	0.048–3.7	<1	<1
Type of karst	Karst low mountain and gorge	Peak forest	Peak mountainous and peak depression	Karst moderate mountain and gorge	Karst low mountain gorge

stone outcrops in these areas, allogenic water rising on non-limestone outcrops is important in the formation of both the surface and the underground landforms. Some particularly large caves are associated with the corrosion caused by allogenic streams. One of the largest caves is Wanhuanyan cave in Chenzhou, Hunan, over 2000 m long and formed by water flowing off granite hills; boulders of granite, over 1 m in diameter litter the cave passages. Low or "degenerate" cone karst occurs; much of this area could be crypto-karstic or exhumed as the Mesozoic red beds have been relatively recently removed (Zhang 1980a; Ren, et al. 1982).

7.3 Karst in the Lower Changjiang Region

This area is characterized by folded limestones of the NE–SW Yanshanian trend in Zhejiang province. SE Hangzhou is an historically well-known area and has been explored since the Sui-Tang Dynasties (581–907 A.D.). The area contains several caves, well known to the poets. The present climate is still subtropical – average annual temperature is about 16 °C and the mean annual precipitation about 1500 mm. The limestones occur in synclines and are Middle and Upper Carboniferous in age and the lower Permian Qixia group. The Chuanshan and Huanglong limestones (Carboniferous) are the main karst formations, but the Triassic Tinglong is also important. Moderately folded country also occurs in Jiangsu, east of Nanjing and south of the Changjiang. Here, the Huanglong limestones (Carboniferous), the Qixia (Permian) and the Tinglong (Triassic) limestones are involved. Folding is NE-SW (Yanshanian). The low hills are generally formed in Devonian sandstones – the limestones occurring in the valleys and on the slopes of the hills.

The limestone landscape in Jiangsu, near Yixing and Daihu (lake) has few of the features of the fengcong or fenglin of the areas to the west and south. It is a fluvial landscape with river-formed caves and solutional and collapse dolines forming where the limestones outcrop or are thinly covered by sandstones; dry valleys in the limestones also occur. Solution rates of the limestones are high and this may be why the Devonian sandstones form the higher ground; it is a well-watered and well-vegetated area. In addition to a slightly cooler climate, the absence of fengcong could be the result of only relatively slight uplift – the hills only being a few hundred metres above the Changjiang. Several river caves occur in the neighbourhood of Yixing, the best known being the Shanjuan cave (Bao Haosheng et al. 1991). The Shanjuan cave is about 2000 m long and has three levels, probably related to the local river levels. The upper level has an atmosphere with a CO_2 content of 0.45%; up to 10000 tourists a day pass through Shanjuan cave and it has been proposed to reduce these numbers. Other similar river caves in the Yixing area include the Muli and Seshi caves. All the caves were filled with weathered residual and riverborne sediments – many of these were removed during the preparations to admit tourists. The caves are still active river caves and contain relatively few speleothems. In the middle level of Shanjuan cave, stalagmitic formations of under 30 000 B.P. have been dated. Some of the sediments in Muli cave contain fossils (panda bear) of middle to late Pleistocene age, about 300 000 B.P.

The Huanglong (Carboniferous) and the Qixia (Permian) limestones in the Yixing area are predominantly micritic – different from the coarse sparry limestones in Guangxi and Guizhou. The individual beds can be from 0.5 to 1 m thick. The Tinglong limestones (Triassic) are mostly thinly bedded. The limestones are covered by yellowish/brown residual deposits from 1–2 m thick; these deposits must represent a long weathering period, but have not so far been studied. Subsoil solution of the limestones is important and weathered blocks are excavated for ornamental stones.

In Zhejiang province nearer to Hangzhou and just to the south of the Yixing district the limestones form the hilly areas. These hilly areas might be regarded as "degenerate cone karst" (Zhang 1980a). This region is also a fluvial karst with probably a long history. It is rather different from the area in Jiangsu discussed above, possibly as a result of a more active tectonic environment.

Pleistocene gravels and red weathered soils occur and there is every indication that during part of the Pleistocene the climate was hotter and more humid than at present. Several well-known caves are found in this part of Zhejiang. Near Hangzhou is the Lingshan cave, also known in the Tang Dynasty. This cave is associated with the river Qiantang and has a fine development of massive stalagmites, which to a climatic geomorphologist would be regarded as having been formed in a hot, wet climate.

The most famous cave is the Yaolin cave which is partly related to a series of river levels (Lin Jinshu et al. 1987). The cave consists of a series of large chambers and is over 1000 m long. There are four cave levels, the lowest level being the present underground stream, at about 28–33 m above present sea level. Lin Jinshu et al. believe that the cave developed from at least middle Pleistocene. Speleothems have been dated by uranium/thorium series and by ^{14}C methods. The oldest speleothems are of middle Pleistocene age (Zhou Xuansen 1981); some of these are over 18 m high. Lin Jinshu and his fellow workers examined the clastic cave deposits of Yaolin; these show a fluctuating base level and variations in the velocity and volume of the cave streams. The cave is multistage, which originated as a phreatic conduit and was only later broken into by the surface rivers as the base levels lowered. Thus, the pebbles and sand deposits found in the lower cave show much erosion and attrition by mechanical ground flow collisons, which are revealed in SEM analysis. Furthermore, there was substantial collapse of the underground passages in the later stages, giving rise to large cave chambers. The cave is now a tourist attraction.

Zhang Shou-Yue and Zhao Shushen studied speleothem ages using uranium/thorium series in other caves of East China. Episodes of speleothem deposition are attributed to the interglacials and non-depositional episodes to the glacials. Hence, periods of speleothem deposition occur (1) from 20 ka to the present; (2) 120–75 ka; (3) 230–170 ka; and (4) 310–250 ka. These periods correlate with high sea levels, and warm and wet climates as observed on the isotopic record of the marine foraminifera (Lin Jinshu et al. 1982; Zhao Shushen et al. 1988).

7.4 Conclusions on the Karst of South China

It will be seen, therefore, that there are many other karst areas in China, south of the Changjiang, in addition to the much studied regions of Guilin, Guizhou and Yunnan. The factors which give rise to the differences of these widespread areas have not yet been studied. Climatic factors are clearly important, but lithology of the limestones and the tectonic environment must also be considered.

The climate of southern and central China, south of the Changjiang, is generally subtropical and tropical. Much of the karst in this area is characterized by fengcong (peak cluster) and fenglin (peak forest), i.e. by karst with distinctive peak forms. Almost all the peak-cluster peaks are conical (cone karst), whilst the peak-forest peaks are generally tower-like (tower karst). Yuan Daoxian et al. (1991) gives a much greater areal extent to these karsts in South China than the maps by Zhao Songqiao (1986) or Zhang Zhigan (1980a). Yuan Daoxian possibly includes some non-peak karst. The peaks result from the deep fluvial dissection of hard fractured Palaeozoic limestones; this dissection accompanied the development of closed depressions and some of the largest cave conduits in the world. Neotectonics also played a considerable part.

However, there is much karst in South China which is not of the fengcong and fenglin types. Subjacent or crypto-karst is important in areas where the limestones emerge from the Eocene red beds (as in Lunan and the Hengyang basin). Nearer the Changjiang valley, some of the karst might be called degenerate peak forest (Zhang Zhigan 1980a), but most of the karst in Jiangsu and Zhejiang is different from the fengcong and fenglin – a result of quite a different lithology, an absence of neotectonism, and a much more fluvial geomorphology.

8 The Karsts of North China

8.1 Introduction

North of Changjiang (the Yangtse), the landscape is part of North China. Limestones are important in this landscape and in the economy, but are often covered or buried; their outcrops are much more isolated than in S China. Limestones outcrop widely in North China (Fig. 50), though not as continuously as in the south. The most continuous area is in Shanxi province where the Cambro-Ordovician limestones form an extensive plateau, but are covered by loess deposits. The Taihang range is the eastern part of the Shanxi plateau and is also in Ordovician limestones. In Shandong there are wide outcrops of mid-upper Cambrian and mid-lower Ordovician limestones in the mid-south of the province. In the mountains to the west and north of Beijing, limestones of various ages are prominent in different places. These include first, the area of Zhoukoudian which is a small area of Ordovician limestones at Longgushan, and which contains the cave of Peking Man (*Homo erectus Pekingensis*). And secondly, in the Xishan along the Juma river where the hills are as high as 1000 m and the karst is in siliceous dolomites of the Wumishan formation, late Precambrian or Middle Proterozoic; the well-known Yunshui cave is in the dolomites of the Wumishan formation. NE of Beijing, near Tangshan there are quite a number of limestone hills, mostly in the Sinian and Proterozoics; many of these hills are being actively quarried for the cement and ceramic industries of Tangshan. Karst limestones also occur in the northern provinces of Heilongjiang, Jilin and Liaoning. In Heilongjiang there are marbles of the early Palaeozoic; in Liaoning in the Taizi river basin, the limestones are Cambro-Ordovician (location map, Fig. 2).

In general, as was seen in Chapter 1, the limestones of N China are older than those in the south (Zhang 1980a,b). They are also quite different in texture and structure. The Proterozoic and the Majiagou Ordovician limestones are more thinly bedded and more micritic in texture with many stylolites, compared with the massively bedded and crystalline sparry rocks in S China; the rocks on the N China platform are of a very different style from those on the Yangtse and South China platforms (Fig. 7). The Ordovician and the Proterozoic limestones (in particular the Wumishan) are very susceptible to physical weathering, particularly frost weathering; Weng Jingtao reports that they are "very fragile" (Liu Zhaihua 1991). This susceptibility to frost weathering is important in view of the Pleistocene climates in N China.

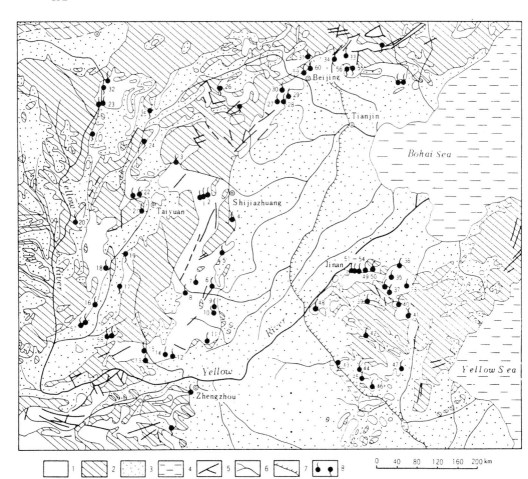

Fig. 50. Carbonate rocks and the distribution of major karst springs in North China. Key: *1* Carbonate rock; *2* insoluble rock; *3* loose deposit; *4* lake and ocean; *5* fault; *6* river; *7* canal; *8* karst spring and its number. Springs in map:*1* Niangziguan S.; *2* Weizhou S.; *3* Pingshan S.; *4* Shigu S.; *5* Baiquan S. in Xingtai; *6* Heilongdong S.; *7* Dongfenghu S.; *8* Xin'ancun S.; *9* Zhenzhu S.; *10* Xiaonanhai S.; *11* Baiquan S. in Huixian County; *12* Jiuli S.; *13* Jinan S.; *14* Sangu S.; *15* Mashan S.; *16* Longzisi S.; *17* Guangshengsi S.; *18* Guozhuang S.; *19* Hongshan S.; *20* Liulin S.; *21* Jinci S.; *22* Lancun S.; *23* Tianqiao S.; *24* Xiama S.; *25* Shento S.; *26* Shuishentang S.; *27* Gaozhuang S.; *28* Ganchi S.; *29* Heiniushui S.; *30* Hebel S.; *31* Zhenzhu S. in Huairou County; *32* Laoniuwan S.; *33* Huangcaowa S.; *34* Shentou S. in Boshan County: *35* Weitouhe S.; *36* Bengshui S.; *37* Longwan S.; *38* Donglongwan S.; *39* Guoniang S.; *40* Nanyu S.; *41* Tongjing S.; *42* Quanlin S.; *43* Yuanyuan S.; *44* Jingquan S.; *45* Yangzhuang S.; *46* Shili S.; *47* Linyi S.; *48* Shuyuan S.; *49* Dongmawan S.; *50* Ximawan S.; *51* Baotu S.; *52* Heihu S.; *53* Zhenzhu S.; *54* Wulongtan S.; *55* Gongyating S.; *56* Dongquan S. in Jixian County: *57* Gudui Hot s.; *58* Haitou Hot S.; *59* Heilongtan S.; *60* Yuquanshan S. (Yuan Daoxian et al. 1991)

Introduction

The limestones in N China form part of the N China massif. There is considerable vertical variation in both the Proterozoic and the Ordovician limestones, but less lateral variation, as a consequence of their being platform limestones. Dolomites and gypsum beds are also important in these older sequences. Structures on the N China paraplatform tend to be block-faulted rather than the close-set folds of the southern Yangtse paraplatform (Fig. 3). The limestones in N China, therefore, tend to be gently dipping over large areas which provide the geological setting for the large hydrological basins and large springs which characterize the karst in N China (Fig. 50).

The most southerly part of N China is the Qinling mountains. These E-W trending mountains have always been regarded as a transitional zone and this is their role also in the karst geomorphology of China. South of the Qinling the present-day climate is generally wet with at least 800–1000 mm precipitation annually; it is much influenced by the SE monsoon with most rain falling in the summer. North of the Qinling the climate is drier, to the NW in Shaanxi substantially so, with an annual rainfall of 150–200 mm; to the NE in Shandong the precipitation is 500–700 mm. Both these provinces could be regarded as subhumid. Much further north, to the NE of Beijing, in the provinces of Jilin, Liaoning and Heilongjiang, the precipitation is again higher, being 800–1000 mm/year; the climate is cooler and the karst quite different. The Qinling line also seems to have been a somewhat similar boundary in the Quaternary. The Quaternary climates in S China may not have been greatly different from those of today – at times much drier and cooler and at other times much wetter, when the great Quaternary mud flows were active (Derbyshire 1983). The evidence for frost action is extremely limited in S China, except possibly at the summit of the Lushan (1428 m) just south of the Changjiang, where Quaternary permafrost has been identified. As we have already seen, there is little evidence for any substantial frost action in Guangxi or even Guizhou except at high levels. However, N China, N of the Qingling line, was very cold and dry in the Quaternary. No ice cap formed in this part of eastern Asia, but the outblowing winds from the interior not only helped to give intense frost but also carried the well-known loess deposits into N and NE China, as in Shanxi. Thus, much of the limestone area in N China is covered by loess and Quaternary scree deposits are also important. Shi Yafeng believes that in the glacial period climate in N China, it was at least 12 °C cooler than at present; the Quaternary permafrost line was close to Beijing (Shi Yafeng 1992).

The Tertiary climates in both N and S China were probably warm and humid and allowed the development of subtropical karst in the Tertiary in both areas. As a result of the continued corrosion and erosion during the Quaternary, and because of the tectonic uplifts, much of the Tertiary karst in S China was destroyed or is only found in small areas of the plateau. Because of the deposition of the loess in N China and possibly because of the dry Quaternary and present-day dry climate, Tertiary palaeokarst is well preserved in N China. This explains many of the differences between N and S

China. The karsts of N China are thus less related to their present processes than those of the south and are more palaeokarstic.

Before discussing the main areas of karst in N China, the limestone areas of the Qinling mountains will be considered. These are regarded by Yuan Daoxian as the most northerly of the S China type of karstlands (Yuan Daoxian 1991).

8.2 Karst in the Qinling Mountains

The Qinling mountains are a Palaeozoic fold belt which suffered severe denudation; they were then uplifted, by about 2000–3000 m, in the Tertiary Himalayan period. The northern slope of the range is steep, overlooking the plain of Xi'an and the Weihe valley by 2500–3000 m. However, the southern slopes are more gentle. The highest parts of the Qinling are over 3600 m. The Qinling form the most significant climatic barrier in China. Apart from forming a rainfall divide as already indicated, they hold back the northerly winds in winter and the SE monsoon in summer. The present-day temperature is annually 18.4 °C, but 12–13 °C in the valleys. Considerable thickness of carbonate rock, predominantly late Palaeozoic, occurs in the mid south-Qinling area (Fig. 51). Although some thousands of metres of limestones were laid down, because of the acute folding and facies variation, the sequences are not as thick as those in S China. Limestones and dolomites of Cambrian, Ordovician, Permian and Triassic age occur. The trend of the strike is E–W. As a result of the folding, the limestones were affected to some extent by metamorphism, as for example in the pink limestones or "marble" of Xigou.

Some areas of limestones in the Qinling have now been studied, particularly those around Zheng'an in the mid-south Qinling. The limestones are Carboniferous and Permian. Here, there is an area of cone karst resembling the peak clusters or fengcong of Guizhou. The slopes of the cones are relatively steep, from 45°–50° and the intervening dolines well developed, up to

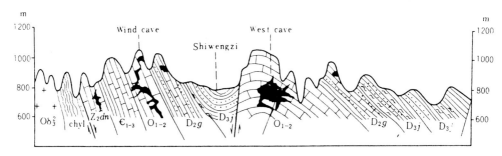

Fig. 51. Geologic profile of Shiwengzi area, Zhashui County, Qingling; *D* Devonian; *O* Ordovician; *Oβ-Z-ε* Precambrian-Cambrian. (After Gao Shiliang 1986)

300 m deep, particularly near the village of Heigou (Yuan Daoxian 1991). There are numerous big springs, such as Yudong spring, which have a high coefficient of variation between winter and summer outflow – up to 100 times. This variation in spring discharge reflects the strong seasonal and monsoonal climate, as in S China. Epikarstic springs are also well developed, as in Guangxi and Guizhou; the epikarstic springs dissolve large quantities CO_2 from soil air and the CO_2 content of the soil in autumn may be as much as 7000 ppm. Epikarstic water is regarded as very aggressive with a high carbonate hardness and probably plays a role in the development of the doline depressions. The underground stream water is less aggressive and softer. The epikarstic water is part of the fissure flow, while the underground streams are part of the conduit flow. The temperature of the epikarstic water is 12.7 °C and 11.6 °C for the underground streams. In the middle Ordovician limestones in the area near Zashui, some quite good caves related to tectonic structures are found. These are sometimes on two or three levels as in Tiandong cave. These levels are regarded as reflecting the stages of neotectonic uplift and are high above the Qianyou river. Tiandong cave has also some fine speleothems as yet undated. These caves at Zashui are believed to have been formed by the allogenic water coming from non-carbonate rocks (granites etc.) of the central Qinling to the north. Allogenic water is also considered to be responsible for some imperfectly developed tower karst along the valley of the Qianyou river.

Yuan Daoxian regards the Zeng'an and Zashui areas as being the most northerly development of peak cluster in China. The latitude is 33° 20'–40' N. If it is assumed that peak cluster formation is dependent upon rapid uplift and relatively high intensive seasonal rainfall together with pure limestone development, then these conditions are met with in the central Qinling. In particular, the rainfall in certain stages of the Quaternary could have been higher than at present.

Tufa deposition is also important in the Qinling and it is still forming. The deposition process is believed to have begun 8303 years ago (^{14}C-dating) and it is thought that certain phases of the Holocene were particularly favourable to tufa formation. Degassing of CO_2 is regarded as important in tufa deposition, but Liu Zhaihua (1991) also stresses the influence of bryophytes and other vegetation.

8.3 The Shanxi Plateau

Shanxi province is an upland province W of the N China plain and E of the upper Huanghe (the Yellow river). The Shanxi plateau is most famous for its loess deposits. Bare karst totals about 33 000 km² – about 21% of the area of the Shanxi plateau, while buried karst is about 102 000 km², 65% of the total area (Fig. 52). The average elevation is more than 1000 m; loess deposits up to 10–30 m thick, cover much of the limestone outcrop. The average annual

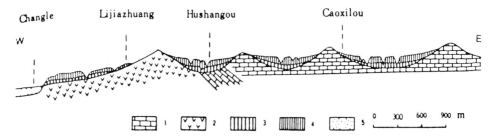

Fig. 52. Ancient karst basins in Shanxi. *1* Limestone; *2* basalt; *3* reddish-brown clayey soil; *4* loess; *5* sand. (After Liu Dongsheng and Zhang Zhigan 1980)

rainfall varies between 400 and 600 mm, decreasing from SE to NW. The low rainfall together with the alkaline loess cover means that modern surface karst is poorly developed. Solutional fissures are well developed however, and 15–25% of the precipitation percolates into the rock, causing subsurface karst features to form. The karst area in Shanxi is seamed with dry valleys, probably the relics of a more humid climate in the Tertiary and Quaternary. Some Chinese geomorphologists regard the dry valley network as an adaptation to modern climatic conditions (Zhang Zhigan 1980a). A further characteristic of the Shanxi area is the occurrence of large springs (Fig. 50).

The total thickness of the Ordovician limestones in the Shanxi plateau is 600–1200 m, being thicker in the north. As suggested earlier, the facies vary greatly in vertical section, but not so much horizontally. The pure limestones are interdigitated with argillaceous carbonates and dolomites, which brings about a multilayered karst. The middle Ordovician which is 400–600 m thick also includes argillaceous dolomites and gypsum bands. The gypsum-bearing beds are usually at the top of each formation; orginally, the deposit was a lagoonal micritic dolomite and argillaceous carbonate with rock gypsum and anhydrite. Preserved beds of gypsum are now found only in deep boreholes; data from several gypsum mines in the Taiyuan area show that in the Fengfeng formation, the gypsum-bearing strata are up to 100–150 m thick, with the actual individual gypsum beds up to about 50 m thick. Dissolution of the gypsum is an important factor in the karst processes of Shanxi (see below).

In the Taiyuan area, over 4500 km^2 of Cambrian and Ordovician limestones are exposed. A study by Wang Fakung and Ma Fengshan (1989) showed that the porosity of the carbonate rocks is related to the lithology. Porosity increases with increasing dolomitic content. Porosity also varies with the type of limestone (micrite, crystalline etc.); generally, coarser grained rocks appear less porous. Dolomites were less soluble than the pure limestones. As a result of these variations and the well-developed vertical variation of the strata, the multilayered karst development is formed. Layers with caves occur at altitudes of 1300–1500 and 900–1000 m. The reasons for this layering are not only related to the vertical lithological variations, but also to tectonics and the Tertiary and Quaternary history.

Seen from the Hebei plain to the east, in the neighbourhood of

The Shanxi Plateau

Photo 24. Edge of the Shanxi plateau, near Niangziguan

Zhengding, the Shanxi plateau shows a steeply faulted scarp face; the scarp is unglaciated, resembling limestone scarp faces in the Causses of S France, (Photo 24). The scarp is dissected by sharp gullies, both fluvial and nival with much scree. Loess deposits are often incorporated into the scree. The scarp is surmounted by conical hills of the old Tertiary karst. The plateau surface is cut by dry valleys (Fig. 53) which are not related to modern rainfall conditions and are most likely to be the relics of Quaternary palaeohydrological conditions. The dry valleys are utilized by modern drainage, particularly subsurface. No work has yet been done on these dry valleys and no map exists, to the author's knowledge, of their distribution (Zhang Zhigan 1980a; Yuan Daoxian et al. 1991).

The long profiles of the dry valleys are of low gradient, about 0.1%. According to Zhang Zhigan, the accumulation of deposits in these valleys is mainly fluvial and is from 20–100 m thick; the deposits may be of Malan age (Q3) (Zhang Zhonghu 1991) and Holocene. The uneven rainfall in Shanxi, 500–600 mm/year falls mostly between late June and September. The dry valleys may flood and be full of water during the rainy season, but the flow is not sufficient to remove the detritus and debris in the valleys out of the region. In the Taihang mountains, to the E of the Shanxi plateau, rivers with a drainage area of a few hundred km^2 have a peak flood discharge of 2000–3000 m^3/s. This will drop to a few cubic metres per second after a few days. The wide alluviated valleys form areas of farmland, known as *Chuans*. Chuans may have catchment areas from a few tens of km^2 to hundreds of km^2 and are 100–

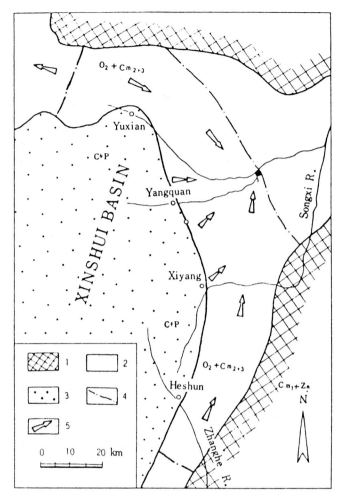

Fig. 53. The karst water basin at Niangziguan, Shanxi. *1* Lower Cambrian-Sinian clastic rocks; *2* Middle Cambrian-Ordovician soluble rocks; *3* Permo-carboniferous clastic rocks; *4* watershed of underground karst water; *5* direction of karst water flow

500 m wide. The slopes are mantled in loess and fine limestone scree 2–10 cm in diameter. If excessive rainfall occurs, the vegetation on the slopes may be broken, the weathered mantle slides down and debris slides take place. In 1963, during a severe flood, large slides of up to 500 m long were formed. The fissures in the limestones are often coated with calcareous tufa (more than 1 mm thick) making the permeability of the limestones very poor. The calcareous deposition also indicates a former and possibly a present phase of carbonate precipitation.

At the present time, the winter temperatures in Shanxi are below $-10\,°C$, so frost action should be important today as well as it was in the Quaternary. Because of the relatively low rainfall, the corrosion rate of the limestones is low, about 10 mm/1000 years, compared with 80–120 mm/1000 years in central Guangxi. However, the rate of mechanical breakdown of the limestones at the present time is considerable, approx, 300–1000 mm/1000 years; this figure is over 30 times higher than that for carbonate corrosion. Solution forms, such as karren, on the limestone surface, are rare, and individual karstic features, dolines, shafts and caves are difficult to find (Zhang 1980a).

Many of the solutional landforms have been buried by the loess and the mechanically weathered debris. Much of the karst relief is infilled with loess, which can fill in hollows over 30 m deep (Fig. 52). The most important solutional landforms of the Shanxi plateau are palaeokarstic. Since the Mesozoic, the upper Palaeozoic strata above the limestones have been stripped off and the present Shanxi plateau was uplifted in the late Cretaceous and then again strongly uplifted in the Pliocene as a unit. During this phase, it is believed that the later Tertiary climate was warm and humid; this was the period of the Yanshanian-Himalayan stage of karstification according to Zhang (1980a). What looks like a peak cluster cone karst is developed on the Shanxi plateau, and is now a relict karst on the plateau surface. The climatic conditions during the Quaternary were not favourable for the development of either surface or underground karst. Closed depressions and caves formed in the Tertiary were buried under the loess and other physically weathered deposits. Several high-level cave systems occur on the Shanxi plateau; these are of Tertiary age and are now dry and are relics of the palaeokarst. According to Chinese hydrogeologists, modern groundwater formed in new fracture systems related to the present conditions; the present hydrology is not related to the old caves, which are part of the aeration zone. There are some small caves that are the result of the present hydrological system. According to Zhang, high-level karst depressions, resembling poljes occur on the Shanxi plateau, particularly at levels above 1000 m. At least three planation surfaces of Tertiary age are believed to occur on the plateau (Zhang 1980).

The most important karstic characteristic in Shanxi is the presence of the large springs and their karst water basins; there are more than 50 big springs with a discharge greater than 1 m^3/s. The maximum discharge of Niangziguang is 17.3 m^3/s and the minimum is 8.1 m^3/s. The Niangziguang spring and the Xinancun spring each have discharges of more than 10 m^3/s and are two of the largest springs in N China. The geomorphological setting of Niangziguang is on the E side of the Shanxi plateau and on the edge of the Taihang mountains (Yuan Daoxian et al. 1991).

The Niangziguang spring occurs in a deeply cut dry valley, the Mianhe, and like other springs in this part of N China has many outlets (Fig. 54). It consists mainly of 11 springs, the most stable and with the largest discharge is the Wulong spring about 1.8–2.0 m^3/s. The area of the whole system is 4667 km^2 of which 2118 km^2 is exposed limestone, the rest is covered karst. Like most of the big springs in N China, it is associated with the edge of a Tertiary down-faulted basin and near the margin of an incised canyon. The size of the catchment area reflects the relatively simple geological structure (Fig. 53). A feature of the springs like Niangziguang is the efficient gathering

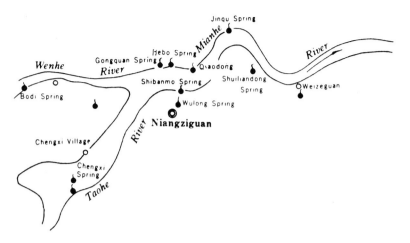

Fig. 54. Location of the main springs at Niangziguan

together of subsurface water over a large area; they may be regarded as huge drainage wells, affecting the whole karst reservoir. The springs are fed by fissures and fractures and *not* by cave conduits and subterranean rivers as are the springs in S China; but strong seepage flow zones occur. Though the flow volume is great, the flow *velocity* is small, about 8.8–23.9 m/day in Niangziguang. Around the spring outlets both today and in the past, there is much tufa deposition. Along the outlets of the Niangziguang springs, tufa deposition is 40 m thick, forming two terraces. The tufa contains plant remains and pollen has been dated by ^{14}C. The lower tufa terrace, the younger, is considered to be late Quaternary in age, while the upper terrace is thought to be mid-Quaternary. As a result of overpumping in the last few years, many of the springs at Niangziguang are now dry (Zhang Zemin 1988).

The regime of springs like Niangziguang is relatively stable compared with the regime of springs in S China; in the south, the springs rapidly reflect recent floods and rainfall. The coefficient of variation for Niangziguang is between 1.1 and 2. The yearly fluctuation of Niangziguang karst water basin does not exceed 20–30 m and the daily fluctuation is no more than 1 m. There is a high water flow in the spring from the recharge of surface water in the rainy season and a later high water flow when the recharge is from the groundwater – about 3–4 months after the rainy season (Fig. 55). Many spring discharges are related to the rainfall of the same year, but Niangziguang and most large springs in Shanxi correlate most closely to a 4–5 year period. The water of the Niangziguang springs was tested for its age using tritium units and it is believed to be about 80 years old. Because of the presence of gypsum in the Ordovician limestones, the Niangziguang waters are of the H_2CO_3–SO_4–Ca type. The total dissolved solids varies from 100–500 mg/l. Corrosion rates are given in *The Karst of China* (Yuan Daoxian et al. 1991).

The Niangziguang spring illustrates the simpler karst hydrological systems

The Shanxi Plateau

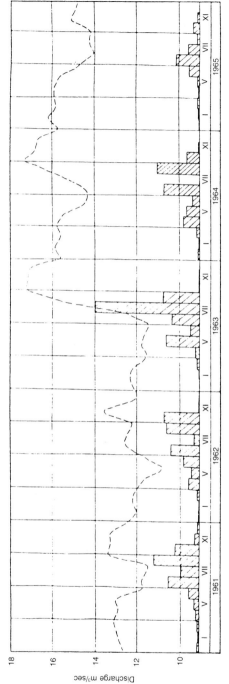

Fig. 55. Hydrograph of discharge of the Niangziguan springs; *dotted line* rainfall (Zhang Zhigan 1980b)

that are more typical of N China – huge systems that are formed in association with the fault blocks of the stable N China platform with thick continuous carbonates and with karst fissure networks as the major water-bearing media. Because of the presence of the gypsum beds, the karst processes are greatly enhanced (Zhang 1980a,b).

The effects of the gypsum beds in the Ordovician Majiagou limestones are very important. Much of the gypsum is in the form of anhydrite. As a result of tectonic movements, the anhydrite forms dome-like structures, as in salt domes, which weaken the surrounding and overlying strata. The percolation of water into the gypsum and anhydrite beds, particularly into the areas of "salt-domed" anhydrite, has two main effects. First, anhydrite expands when hydrated to gypsum by 67%; such an expansion causes the limestone beds in association with the gypsum beds to become broken and disordered. Secondly, percolating water dissolves the gypsum before the limestones; the dissolution rate of gypsum is five to ten times higher than the dissolution rate of limestone or dolomite. As a result of these two effects, crushed gypsum-bearing layers and gypsum breccias resulting from gypsum dissolution are quite common in the Majiagou limestones.

Accelerated solution of the gypsum causes the development of large underground cavities; the already broken strata, as described above, overlying the gypsum beds, collapse and subside and roof falls occur. This process leads to palaeokarstic collapse columns, which are a feature of the middle Ordovician beds; such columns are usually over 100 m deep, as illustrated in Fig. 56. The incidence of collapse columns was increased by the mining operations in the removal of coal (Permo-Carboniferous), which lie above the Ordovician limestones in parts of Shanxi; as a result of coal extraction, much water penetrated into the Ordovician limestones, which induced further collapses. Sudden water gushes took place, with disastrous effects on the mines. In parts of Huoxian county, there is an average of 37 collapses in an area of 1 km^2, and in one area this figure rises to 72/km^2. Since the collapse takes place at depth many of the palaeokarstic collapse columns have no surface indication and are hidden (Yuan Daoxian et al. 1991).

The Taihang range lies on the eastern edge of the Shanxi plateau and is transitional between the plateau and the N China plain, both in its tectonics and its geomorphology. The relief is high, 1400–2000 m, with steep slopes. The mid-Ordovician carbonate rocks are much more faulted than those on the Shanxi plateau and have been cut into differently sized fault blocks. As a result of the intense faulting, the whole mid-Ordovician series, 400–600 m thick, has become hydrologically integrated and forms a large united water-containing mass. As in Shanxi, old planation surfaces exist and bauxite material is said to exist in the old (Tertiary?) karst depressions. Studies in the water resources of the Taiyuan area were made by Zhu and by Zhang Dachang (Zhu Fengjun 1987; Zhang Dachang 1988).

The hydrological regime of the Taihang range is more dynamic than in Shanxi. The spring peak flow in the Taihang lags behind the rainfall peak by only 2 months; in the Shanxi plateau the spring discharge curves rise and fall

The Karst of Shandong

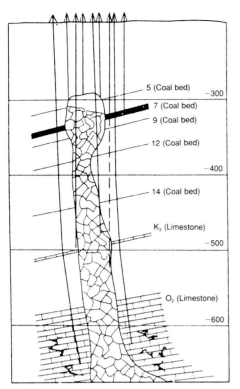

Fig. 56. Karst collapse column in Fangezhuang Coal Mine, Kailuan, Shanxi. (Yuan Daoxian et al. 1991)

more slowly and there is a time lag of 4–6 months behind the rainfall peak. In the Taihang mountains, there are two hydrodynamic systems. Firstly, there is an upper one which has a circulation depth of about 600–800 m and which contains 80–85% of the total recharge and is the more important in the water resources of the area. Secondly, there is a lower system in which the groundwater is more confined and where the discharge is largely into the Quaternary rocks of the N China plain; the circulation depth of this system can reach several thousand metres.

8.4 The Karst of Shandong

Shandong is in E China, the peninsula between the Yellow and Bohai seas (Figs. 2 and 50). Over 1100 m of carbonate rocks occur in mid-Shandong. The limestones are generally thin-bedded micritic limestones; they are the lower middle Cambrian, the Wormkalk limestone of the upper Cambrian, and the micritic dolomites of the middle Ordovician (Photo 25). The limestones occur in a monocline, dipping off the metamorphic rocks on the southern slopes of Lushan (Shandong), 1108 m; dips are between 10 and 15°. The dolomitic beds contain many stylolites and are very different from the limestones in S China. Shandong was uplifted by about 300 m during the later Pliocene and in the post-Tertiary.

Photo 25. Thinly bedded micritic limestones in Shandong

The present mean annual precipitation in Shandong is about 767 mm, decreasing to the NW. According to Yuan Daoxian in *Karst of China* (1991, p. 104, Table 3.14), the modern corrosion rates in Shandong reflect the present-day precipitation. The more humid area of Xinxue SE of the province has an annual precipitation of 800–900 mm and a corrosion rate of 31.73 m^3 km^{-2} year.$^{-1}$. The drier area of Yufu to the NW has an annual precipitation of 600–700 mm and a corrosion rate of 25.71 m^3 km^{-2} year^{-1}. These are the present-day rates, but as in many other such calculations, they give little idea of the origin of the karst landforms. This is due to the fact that many of the features of the karst in Shandong are palaeokarstic, and are only slightly modified by modern processes. As in Shanxi, the most striking aspects of the karst in Shandong are the old palaeokarst forms on the plateau surface, the long and wide dry valleys and the present-day large springs. Evidence of Quaternary and probably modern frost action is abundant, and scree deposits are often intercalated with loess. Frost and snow gullies dissect the plateau slopes as a result of periglacial conditions in the Quaternary. Even today, the average January temperature in Shandong is −10 °C.

The surface of the Shandong plateau bears a relict, shallow, fluvial cone karst (Photo 26). Three karst denudation surfaces are said to exist on the plateau, believed to have been formed in the Pliocene and early Pleistocene. On one of the surfaces, the Yangping, in the area of Lushan, there are several palaeokarst depressions. Dolines up to 300 m deep occur on the plateau surface at 740 m. They contain a reddish soil that contains sporopollen, indicating a warm palaeoenvironment; the red soil (? terra rossa) has been dated by

Photo 26. Shallow Palace cone karst on the plateau surface in Shandong

thermoluminescence to about 1.2 million B.P. Associated with these old karst depressions are caves and water inlets which are connected with the modern springs; this shows that the hydrological systems are well developed and that the depressions are still the recharge points for the groundwater. In the neighbourhood of the metamorphic and gneissic rocks, allogenic water is important in the present-day continuing karstification. High-level caves formed in the late Tertiary have been uplifted and occur above the modern developing caves at the present groundwater level – an example is the dry Shanhu cave (coral cave) in the Tumen area (Liu Zhaihua 1991).

The fossils (teeth and part of a cranium) of *Homo erectus yiyuanensis* were found in the Lushan area of South Shandong; they were discovered in yellowish clays in a fissure cave in Ordovician dolomites at a low level (only a metre or so) above the present river level. It is believed that the cave fissures may not be the original place of deposition of the fossils and that they were transported and sludged down and redeposited from a middle level cave. The age of this *Homo erectus yiyuanensis* has been dated as middle Pleistocene, the same age as isotopic determinations of calcareous material that is found in the middle level caves. Yiyuan Man is believed to have been contemporaneous with *Homo erectus pekingensis* (Peking Man) and the Heshan Man from Heshan in Anhui province.

Dry valleys lie at the foot of the plateau slopes. The valleys are some kilometres long and at least 100–200 m wide in their transverse profiles. They could not have been formed under the present hydrological conditions. They are now being modified and used as drainage channels by the allogenic streams draining from the higher non-calcareous metamorphic rocks. As the streams pass on to the limestone, they sink slowly into fissures in the rocks and percolate underground. Usually there is no one sinking place, but a succession of leakage points. In some valleys, the underground fissures link up with caves formed in earlier periods. In their upper and middle sections, the longitudinal profiles of the valleys are relatively gentle and their transverse profiles are also fairly flat. They are filled with fluvial and other deposits which so far have not been investigated. In their lower reaches, nearer to the main allogenic rivers, the valleys may have active streams, but this activity is not enough to remove the great slugs of deposits which exist in these valleys. It is likely that these valleys are of Pleistocene age. In humid phases they may have had reasonably large streams, but in colder and drier phases, they became filled with slope deposits, frost debris, and river deposits with which they must be filled today. The valleys present an interesting problem to the palaeohydrologist and resemble to some extent similar valleys in the limestone areas of central Europe (Starkel et al. 1991).

Erosion of the top soil in the Shandong area has made many plateau slopes bare allowing a few karren to develop. In some parts, the removal of soil and vegetation on the gently dipping strata has allowed bare limestone pavement features to form – an example of a limestone pavement being formed in a non-glaciated area.

The other features of the karst in mid-Shandong are the large springs, as in Shanxi; they are particularly important in Jinan and have recently been studied by Xi Devin (1991). Jinan the capital of Shandong province is known as "Spring City". There are about 108 spring outlets in 4 km^2 of the urban area. The most famous are the four groups – Baotu springs (Spouting) shown in Photo 27; Heihu springs (Black Tiger); Zhenzhu spring (Pearl); and Wulongtan spring (Five Dragons pool). The city of Jinan lies north of Mount Taishan and is bounded on the north by a piedmont plain and the Huang He (Yellow river). The basement rocks of the region are the metamorphics of the Archean Taishan group; these are covered by the Cambrian, and lower-middle Ordovician; Quaternary sediments cover much of the northern and southern areas. Mesozoic diorite and gabbro of Yanshanian age (Mesozoic) occur north of the city. The Yellow river is about 8 km to the north of Jinan. The Cambrian is characterized by interbedded limestones and shales, while the Ordovician system is composed of thick-bedded limestones, and dolomitic limestones. These beds dip at about 5°–10° to 20° NE forming a monoclinal structure. The beds are generally without folds but faults are well developed; these trend NNW-SSE or NW-SE. The faults have an important effect on the hydrogeology and cut the Jinan monocline into distinct parts and control the movement of the groundwater and the development of the karst. The aquifer of Jinan lies between the Dongwu fault in the E and the Mashan fault in the W (Xi Devin 1991).

The upper Cambrian and the lower-middle Ordovician limestones are the main aquifers with a thickness of 800–1000 m (Fig. 57). Karst water is impounded in the Ordovician limestones and is abundant but non-uniform

The Karst of Shandong

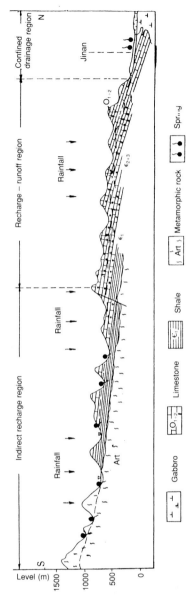

Fig. 57. Groundwater movement in the Jinan area. (After Xi Devin 1991)

Photo 27. The Baotu spring in Jinan city, Shandong

(heterogeneous). In the southern part of the area, the water originates 50–100 m beneath the surface with a small yield. In the piedmont plain, karst water is much more abundant and a single well may yield more than 10 000 m^3/day; or as much as 1.85 m^3/s as in the Baotu springs. Rainfall is the main source of recharge. The annual precipitation in Jinan averages 600–650 mm with 70–80% between July and October. There is a close relationship between rainfall and the springs' discharge. Seepage of surface water is also important in recharge, particularly from the two main rivers, the Yufu and Beisha. The karst water moves from south to north along the dip. In the region of Jinan, this water is blocked by the Mesozoic igneous rocks (gabbro and diorite) where it accumulates at the contact zone between the limestones and the igneous rocks. The NNW trending faults confine the karst water and in the urban district of Jinan, the limestones are surrounded by the igneous rocks. As a result, the water gushes out through fissures in the karst as large springs. The total area of the Jinan catchment is about 1500 km^2.

In recent years with the development of Jinan and the increase in water use in industry and agriculture, the groundwater in the Jinan area has been overextracted; the famous Jinan springs have decreased in volume and now disappear altogether in the dry season (Fig. 58). In particular, the extraction of water from the western part of Jinan has been a main cause in reducing the outflow of the springs since 1982; this is because the area S and SW of the urban district is the main recharge area of the springs. Furthermore, the annual precipitation in the Jinan area in the last decade has been markedly below 600 mm; in 1989, for instance, it was only 308 mm. In the dry years, the

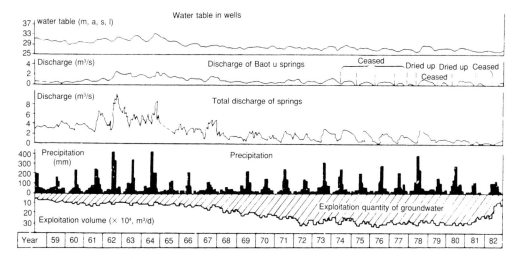

Fig. 58. Hydrograph of karst water in Jinan, Shandong province. (After Chen Zhenping 1988)

springs now disappear completely. In 1990 when precipitation was nearly 1000 mm, the springs recovered their flow (Li Chuanmo 1985, Chen Zhenpeng 1988).

Several projects have been introduced to protect the Jinan springs, especially by controlling the extraction of water. The development of new water resources is clearly necessary. One of the new projects is the suggested pumping of water from the Yellow river. As Xi Devin (1991) reports, the populaton of Jinan is making "every effort to keep the term 'Spring City' corresponding to the facts" (Xi Devin 1991).

8.5 The Karst of the Zhoukoudian and Xishan Areas, SW of Beijing and Other Hilly Areas near Beijing

The Xishan lies about 50 km SW of Beijing and forms a group of low mountains. Carbonate rocks of platform type of middle and upper Proterozoic and lower Palaeozoic make up the main strata. More than 1000 m of siliceous dolomites of the Wumishan formation (Proterozoic) and the Jixian system of the middle Palaeozoic are exposed, in addition to about 450 m of the middle Ordovician Majiagou limestones. The main structure lines of the mountains are the result of the Yanshanian orogeny (Cretaceous) with faults, and joints striking approximately S-N or E-W. There was also differential uplift in the Cenozoic and further uplift in the Pleistocene and up to the present day – the area still being affected by earthquakes. As a result, the rivers cut steeply into the limestones and vertical caves and karst fissures developed. Three terraces of Quaternary age (mid- and late Pleistocene) occur along the Juma river which cuts through the Xishan. High-level caves sometimes of considerable size above the present river levels developed in both the Majiagou limestones (the Shi Hua cave, 200 m) and in the Wumishan dolomites (the Yunshui cave, 500–600 m a.s.l.). Both are mainly horizontal caves and probably formed in warmer climatic phases.

There is abundant evidence for karst planation in the Pliocene and for a warm and humid climate at that time in the Xishan. In the Pleistocene, there were relatively warm and humid phases and cold and dry phases in the glacials. A reconstruction of the palaeoclimates of the area since the Pliocene is given by Weng Jintao (Weng Jintao, in Liu Zaihua 1991). Weng Jintao reports that thick Pliocene calcareous "terra rossa" is found in the caves and on the planation surfaces. Closed dish-like dolines also occur on the surfaces. So the Pliocene was a period of strong karstification.

The Shihuan cave has six layers connected by vertical or inclined caves. It has a great variety of speleothems, including cave shields. The Yunshui cave is also a tourist cave; it consists of six large chambers, forming a bead-like passage. Large stalagmites up to 37 m in height occur, indicating formation in a warm and humid climate. Uranium/thorium dating suggests that the age of the largest stalagmites is between 290 000 and 320 000 B.P.

The caves of Zhoukoudian lie on the edge of the Xishan area 47 km south west of Beijing. In this area, in the Longgushan (or Dragon Bone hill), is the Peking Man's cave – the home of Peking Man (*Homo erectus pekingensis*). The Longgushan is a small hill in Middle Ordovician limestones of the upper Majiagou formation and at about 128 m a.s.l. The strata form a secondary anticline, with wavy bedding planes and numerous drag folds formed by intraformational sliding; the dip is generally NNE, with dip angles varying from $10°–70°$ (Ren Mei et al. 1981). The limestones are well jointed, those in the $10° N–20° E$ direction being the most important in the karst development, and caves and fissures are developed along them. The Majiagou limestones at Longgushan consist of micritic dolomite and micro-crystalline dolomites and dolomitic marble; micro-fissures are well developed. The limy-micritic dolomites (MgO content 18.55%) are easily attacked by physical weathering and large quantities of collapse breccias caused by frost weathering are an important feature of the deposits in the Zhoukoudian caves.

On the Longgushan, there are four main caves which have yielded evidence of occupation of early man (Fig. 59; Liu Zechun 1985). The caves developed along the bedding planes or joints. The most important cave, the Peking Man's cave, is a vertical cave more than 40 m high and formed in highly dipping limestone beds; the steep walls of the cave correspond to the dip of the rocks at $60°–70°$. At least 40 m of deposits have been investigated in Zhoukoudian cave I (Peking Man's cave), but it is believed that there are at least another 40 m of infill that have not yet been examined. Peking Man's cave is the lowest cave in the hill; Upper cave, New cave and Cave 4 are all higher up the hill; the Upper and New caves are more gently inclined than the Peking Man's cave.

During the Pliocene, it is believed that the area suffered fluvial denudation under a moist subtropical climate; this is referred to by Ren Mei et al. (1982)

The Karst of the Zhoukoudian and Xishan Areas   171

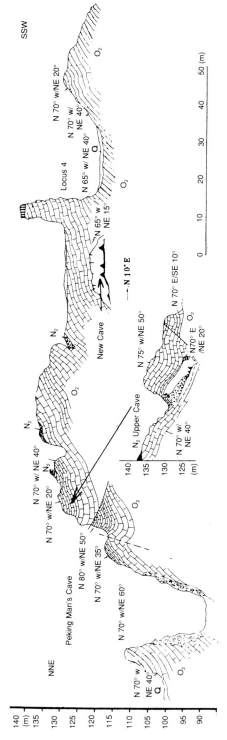

Fig. 59. Geological section of Dragon Bone hill at Zhoukoudian. (Lui Zechun 1985)

as the Tang Xian peneplain stage. It was a time which was favourable to karstification and vertical caves were formed along the steeply dipping limestone strata. In the early Pleistocene, the Tang Xian surface was uplifted and the Peking Man's cave was further corroded and enlarged. The Zhoukouhe (river) downcut further into the limestone; lateral erosion by the river destroyed the eastern wall of the cave, forming a small entrance which is about the site of the present eastern cave entrance. Fluvial deposits of the Zhoukouhe and sheet flow deposits entered the cave and formed a large tract of relatively flat floor. At the same time, the eastern cave entrance was enlarged and this was the way by which Peking Man came into the cave. Stone implements were found from the 13th layer upwards – over 10 000 pieces of stone tools were discovered in the cave. The roof of the cave was not continuous – there was an opening in the top which connected it with the outside and gave air and light. The thick cave deposits in Peking Man's cave are rare occurrences in Chinese caves (Ren Mei et al. 1981). The deposits have been divided into 17 layers; between 1921 and 1929 several hundred pieces of skulls and other bones belonging to *Homo erectus pekinensis* were discovered in layers 11–8, the most important appearing in layer 10 (Ren Mei et al. 1981; Fig. 60). The results of palaeomagnetism show that the deposits below the 14th layer were formed in the Matuyama Reversed epoch and are, therefore, older than 730 000 B.P. Above the 14th layer, the deposits accumulated in the Brunhes Normal epoch for 730 000 years: 34 m of sediment have accumulated since the 14th layer, therefore, most of the sediments have been deposited in the normal Brunhes epoch (Liu Zechun 1985). The site has yielded at least six more or less complete crania in addition to cranial fragments and other humanoid bones and teeth. Besides the humanoid fossils, stoneware, ashes, and contemporaneous animal fossils have also been unearthed. Uranium/thorium series dating of the fossil bones shows that they are more than 300 000 years old (Zhao Shushen et al. 1988, p. 298); dating of ashes in the 10th layer gives a date of 462 000 ± 45 000 B.P. (by fission track dating) and those in the 11th layer of 520 000–610 000 B.P. (by thermoluminescence). The oldest Peking Man bones are usually considered to be about 500 000 B.P. (Liu Tungshen 1988; Liu Zechun 1985). The palaeoclimatic curve in Fig. 60 is linked to the loess-based Quaternary chronology (Liu Tungsheng 1988), as given in Chapter 1 (Fig. 12).

The sediments in the Peking Man's cave can be divided into five groups:

1. Collapse and slide deposits. These are chiefly limestone breccia and are found from the 13th layer upwards, especially in the 8–9th, 6th and 3rd layers. About 50% of the breccias are more than 12 cm long and a few are 2–3 m long. They are mainly the results of the collapse of cave roofs. In the Pigeon chamber of Peking Man's cave, extensive deposits of the cave breccias of the 6th layer occur; from this it is usually assumed that the roof of the Peking Man's cave fell in during the period of the 6th layer. It was probably during this period that activity concentrated on the western part of the cave, as the eastern entrance became blocked.

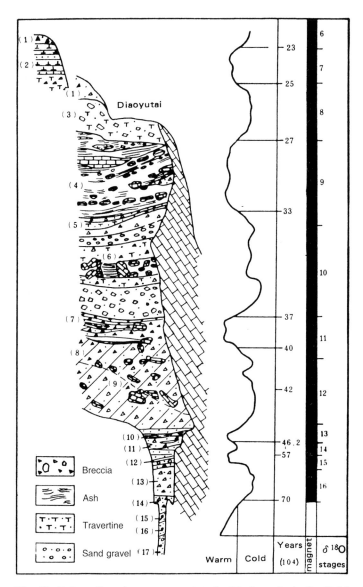

Fig. 60. Palaeoclimatic curve associated with deposits in the Peking Man's cave. (Lui Zechun 1985)

2. Fluvial sediments brought into the cave by the old Zhoukouhe. There are several layers of sand, silt and gravels. The sands of the 17th and 12th layers are similar to the fluvial deposits of the old Zhoukouhe and to the sands of present Zhoukouhe. The palaeomagnetic determinations differentiate between deposits of the old Zhoukouhe and the lower gravels of the present Zhoukouhe. The gravels of the 14th layer came from upper gravels on the top of the cave. A table of the sedimentary characteristics of the cave

deposits and of sediments outside the cave is given in Table 16 (from Ren Mei et al. 1981).
3. Deposits brought into the cave by sheet flow. This kind of cave deposit includes clay, silt, sand and gravels. They are derived from the large amounts of weathered debris which covered the hillslopes. Some of the limestone breccias in the 3rd layer are rather rounded, suggesting corrosion by rain wash.
4. Ash layers. These result from fires and first appear in the 10th layer; they are thickest in the 4th layer where they are up to 6.9 m thick. The distribution of the ash layers gives a valuable key to the reconstruction of the living activities of early man in the cave.
5. Carbonate deposits. Stalactites and thin interbedded carbonate deposits occur in some layers, particularly in layers 8 and 5. They indicate a warm period.

Ren Mei et al. (1981) concluded that the cave deposits cannot in any way be correlated with terraces outside the cave. Deposits from the lowest parts of the cave to Peking Man's cave are older than 700 000 B.P., whereas the lower gravels on the terraces of the Zhoukouho are younger than 700 000 years. As already indicated, gravels from the 14th layer came in from the top (Table 16).

The sequence of deposits in Peking Man's cave have been well studied by many Chinese scientists and the palaeoclimatic events that they suggest, analyzed and correlated with the palaeochronologies derived from the ^{18}O oceanic record and from the loess stratigraphy of N China. This is well discussed by Liu Zechun and the diagram (Fig. 60) is taken from his article (1985, p. 149, Fig. 12). Liu concludes that during the peroid of human occupation of the Peking Man's cave, the surrounding landscape was covered by temperate forest and wooded steppe. During a warm interglacial, temperatures may have been 5 °C higher than the present mean of 11.8 °C; during glacial periods, the temperatures may have been about 10 °C lower than those of today.

It is believed that Peking Man lived in the cave until about 230 000 B.P. At about the time of the deposition of the third – first layers, the cave almost completely filled up, and man left the cave. Peking Man, therefore, lived in the cave continuously for about 300 000 years. The environment was particularly favourable for early man. The location of the cave near the Zhoukouhe gave easy access to water; the cave, in addition to its eastern entrance, had a vertical opening on the top, and there was enough light and air inside the cave; furthermore, the dense temperate forest and wooded steppe at that time yielded abundant wild animals (Ren Mei et al. 1981).

In the caves in the same hill, above the Peking Man's cave are the relics of Shanding Cave Man, who lived about 20 000–18 000 years ago. Zhoukoudian is thus, an important site; the museum attached to the site is now an active centre for archaeological and palaeontological research for the area.

A feature of many of the limestone hills west and north of Beijing is well developed frost pinnacle karst and its accompanying scree deposits. They are

Table 16. Sedimentological characteristics of cave deposits in Peking Man's cave and sediments outside the cave. Data from Ren Mei et al. (1981)

Stratigraphic layer	Heavy mineral assemblage	Clay mineral[a] association	Other lithological characteristics	Chemical composition in quartz grains
1. Cave deposits				
(a) 17th layer	Sphene-magnetite-limonite-hornblendeapatite	Kaolinite, glagerite, illite, vermiculite, epidote	With cross-bedding, rounded small gravels (metamorphic rocks), much mica in sand	Major: Si Trace: Al, Zn
(b) 14th layer (formerly basal gravels)	Sphene-limonite-zircon-magnetite	Illite, kaolinite, vermiculite, epidote	Lithological composition of gravels similar to upper gravels	Major: Si Trace: Fe, Al, Zn Nb, Ta
(c) 12th–13th layer	Magnetite-limonite-apatite-sphenezircon	Illite, kaolinite, vermiculite, epidote	Red-brown clay breccia with coarse sand layers, locally clay forms aggregates	Major: Si Trace: Ba, Ta
2. Sediments outside the cave				
(a) Upper gravels	Sphene-limonite-zircon	Kaolinite, glazerite, illite vermiculite, epidote	Sandstone, schist, slate, phyllite and quartzite	
(b) Gravels at eastern cave entrance	Sphene-magnetite-limonite-hornblende	Kaolinite, glagerite, illite, vermiculite, epidote	Lithological composition of gravels similar to lower gravels, but their diameters are smaller	Major: Si Trace: Al, Zn Nb, Ta
(c) Lower gravels	Sphene-magnetite-hornblende-zircon-limonite-apatite	Illite, kaolinite, vermiculite, epidote	Siltstone, slate, coarse and fine sandstones	

[a] We are indebted to Prof. Xiong Yi, Xu Jiquan and Yang Deyong of Institute of Pedology, Academia Sinica for their help in the analysis and explanation of clay minerals.

largely of Quaternary age and are palaeokarstic. The frost pinnacle karst is particularly well seen in the Xishan along the Juma river gorge, near Shidu. At Shidu, the river cuts through the thinly bedded siliceous dolomites of the Wumishan formation (middle Proterozoic). The dip of the strata is relatively gentle (5°–15°) and vertical joints (in three sets) are strongly developed. These dolomites are very susceptible to frost weathering, but have great compressive strength (Liu 1991). The intense frost action took place along the strong vertical joints; great aprons of scree lie at the foot of the rock columns and frost and snow gullies occur along the joints. Between the joints, vertical upstanding and tapering pinnacles of the dolomite remain (Photo. 28). The summit of the original peneplain is at about 1000 m and the pinnacles tower up to 700 m above the floor of the Juma river. The spacing of the pinnacles is related to the jointing; they are 10–20 m wide and are very "close set". This spacing is of quite a different order from the spacing of peaks in a cone karst which is related to an original drainage system which initiated the cone karst and the regional layer scale fracture systems. A morphometric comparison between the similar (though higher level) pinnacle karst of Lhasa in Tibet with cone karst of Guizhou has recently been done by Zhang Dian (1991) and shows the distinctiveness of the pinnacle karst. Because the Shidu area is basically a frost-formed karst, closed depressions, underground drainage and other solutional phenomena are not widely developed; the mechanical frost action is the most predomant action upon the dolomites, at least in the cold phases of the Quaternary.

The modern winter temperature in the Xishan is below 0 °C and modern frost action is active. However, the most striking feature of the Xishan today is the summer growth of vegetation and the stabilization of the old Quaternary frost forms. The screes and gullies are largely grassed over and the Quaternary forms are "preserved" by the present vegetational growth. Carbon dioxide values in the soils are now quite high and chemical denudation of the dolomites today may be more important than the mechanical denudation; however, modern chemical solution has not yet had the time to change the great impact made by Quaternary periglaciation upon the Xishan.

Similar relict pinnacle frost karst can be found in many other limestone hills in the Beijing area – such as at Badaling near the Great Wall and in limestone hills near Tangshan, NE of Beijing. If these relict pinnacle karsts are compared with the actively forming frost pinnacle karsts in the Minshan of West Sichuan and the even higher limestone mountains outside Lhasa in Tibet (at over 3500 m), it will be seen how the fossil forms in the Beijing area may have originated.

8.6 Karsts in NE China

NE China consists of the provinces of Jilin, Liaoning and Heilongjiang (Figs. 2 and 4). Less is known about these areas but they form a contrast with the

Photo 28. Quaternary pinnacle karst at Shidu, Juma river, shidu near Beijing

more arid areas of Shanxi and Shandong. The rainfall is much more abundant, 800–1000 mm/year, and because the provinces are much further north, evaporation is much less; the climate is, therefore, cool temperate humid, rather than subhumid or subarid. It is believed that the Tertiary climates were humid and warmer than they are today. Apart from the cold periglacial phases of the Quaternary, the karst developed continuously throughout the Tertiary and Quaternary. The karsts are quite different from those in Shanxi and Shandong. They are multilayered, with caves developed at well-defined intervals. In the Taizi area of Liaoning, limestones of Cambro-Ordovician age occupy the central part of the valley; six layers of caves occur along both flanks of the Taizi river, five layers are above the modern base level and one below it. The continuous process of karstification is much more like the conditions in South China; as a result of this continuity, the circulation of karst water is relatively well developed. There are long underground streams; in the Xiejiawaizi underground stream, over 2000 m are accessible by boat, with great halls and water 2–5 m deep. The groundwater is generally unsaturated and the water circulation is very rapid.

The most northerly limestones in China are in the Xiao Hanggan range in Heilongjiang (47°5′N). Early Palaeozoic marbles (292–1773 m thick) are quite karstified. Caves, sink holes and underground streams are found. In this area today frost lasts for 6–7 months of the year, but because of the relatively high rainfall, especially in the forest districts, and because of high humidity and low evaporation, karstification is active. Allogenic water is plentiful (Yuan Daoxian et al. 1991).

The impression gained from work in NE China is that palaeokarst is less important in NE China than it is in the karsts of Shanxi and Shandong, but much work remains to be done.

Limestones occur along the coast in East Liaoning near Dalian. Dolomitic limestones of the Sinian and of the Cambro-Ordovician are exposed in the north along the Bohai Sea and to the south along the Yellow Sea. Springs are abundant and flow along the many faults; submarine springs are also important, and are partly the result of the sea level changes of the Quaternary. The interface between seawater and fresh water is an important zone of mixture corrosion, in which caves are frequently formed; they are similar to the caves formed by mixture corrosion in Yucatan, Central America, and described by Back et al. (1984). In the E Liaoning peninsula, several submarine layers of caves developed due to Quaternary fluctuations in sea level. Six cave layers are believed to exist below sea level, the deepest so far discovered being 120 m below sea level. Peat layers are found in the Yellow Sea; one layer 48–50 m below sea level was dated by ^{14}C to be 12 400 years old, although the karst is older.

8.7 Deep or Buried Block Mountain Karst in N China

As a result of oil exploration, many block mountains have been discovered beneath the Cenozoic plains of NE China. The buried mountains are faulted blocks shaped by differential block movement and denudation. Important oil and gas reservoirs and large amounts of geothermal hot water occur in these blocks. The reservoirs are in the Proterozoic and Palaeozoic carbonate strata. The area most closely studied is the Bohai gulf.

The Bohai gulf oil and gas area includes the Cenozoic subsidence zone between the Taihang and Yanshan mountains, Liaoning peninsula, and the central – southern Shandong mountains. The Cenozoic deposits vary from 1 to 10 km in thickness.

The karst in the burial block mountains formed in the subsidence areas. The karst in the block-faulted areas of Yunnan (previously described) are karsts in areas of block faulting that have been uplifted. However, in the Bohai area, the formation of the buried block mountains involved two stages, in addition to burial and subsidence (Yuan Daoxian et al. 1991):

1. The stage of the block mountain formation; this occurred in Eocene and Oligocene times. Torsional block faulting was characteristic and in most cases blocks rotated as they subsided. By the end of the Oligocene, a series of en échelon block mountains (horsts) and basins (graben) had been produced.
2. The burial stage of the subsided block mountains; at the beginning of the Miocene, the block movements ceased and were replaced by large-scale isostatic depression. By the early Pleistocene, the highest mountains were buried and the mountain summits had been buried up to 4 km in depth. The Palaeogene deposits contain subtropical sporopollen and subtropical plants

with gypsum beds. It is inferred that the Oligocene, in particular, was the main epoch of karst development in the block mountains.

The problem of karstification and permeability of the deep carbonate strata is highly important. The exploration of the oil fields in the Bohai gulf area has given much information. In the buried mountains of Hebei 60% of the bore holes drilled into carbonate rocks have struck caves. In particular, the Wumishan dolomites and the Majiagou limestones are high-yielding oil formations; the Gaoyuzhang and Fujunshan dolomites are also high yielders. The degree of karstification and permeability of the buried limestones in the central Hebei depression are largely similar to those in the surrounding exposed mountains (as in the Taihang and Yanshan mountains). However, compared with the surface karstification, the karstification of the buried Wumishan dolomites is greater than that of the Majiagou limestone (at the surface this condition is reversed; Zhang 1980b). The buried mountains of the Bohai gulf involved three main stages of karstification (in the Caledonian, in the Yanshanian and in Himalayan stages). Strata that have undergone all three stages are characterized by the most intense karstification and the best permeability, so they are the main reservoir formations in this area.

Analysis of data of head pressure, chemical composition and water temperature show that in the area of buried block mountains, each depression usually constitutes a separate hydrogeological unit. From the marginal mountains area to the centre of the depression, there are five zones. The changes in the chemical groundwater composition from the recharge area to the discharge area are shown in Table 17. Most of the high-yielding oilfields are in the pressure area.

The temperature, chemistry and hydrodynamic conditions of the groundwater in the buried carbonate mountains are entirely different from those in the Tertiary aquifers. Compared with the Tertiary aquifers, the permeability of the buried mountains is much higher (10–100 times), the total dissolved solids content 2–10 times lower and the thermal gradient 1–3 times smaller. It is believed that these differences are due to the great permeability of the buried strata (Zhang Zhigan 1980a).

8.8 Conclusions on the Karsts of N China

Thus, the extensive areas of karst in N China form a contrast to the highly dissected "peak" landscapes of S China. This is due to three main factors:

1. There are great differences in the lithology of the older limestones in the north compared with those in the south; they are largely micritic and thinly bedded in the north, but are massive and sparry in the south. These differences alone can cause variations in slope development, in reactions to the weathering agents, and in the speed and methods of penetration of underground water through the rocks. The micritic limestones though being thinner bedded may be more argillaceous and "tighter" and pervent the easy passage of underground water, except in small fissures. This could affect stalagmite and stalac-

Table 17. Chemical groundwater-recharge area to discharge area, Bohai gulf. (Zhang Zhigan 1980a, Table 11, p. 568)

Chemical composition	Recharge area	Intense alternation area	Weak alternation area	Pressure area	Discharge area
Total dissolved solids g/l	0.45	0.45–1.0	1.0–3.0	3–5	5–10
Na/Cl	1.5	1.4–2.0	1.2–1.4	1.0–1.2	1.0–1.2
$\delta_2 O_{18}$	–1.3	–	–1.0	+0.3	+0.9

tite formation, making them slower to develop in the northern caves. Thus, differences in stalagmite and stalactite formation may not, therefore, be caused solely by climatic differences. The importance of the gypsum beds in the Magiagou (Ordovician) limestones in the north cannot be overestimated.

Though in both N and S China, limestones are mainly of the platform type, their later geological history has been quite different. Karst limestones in N China are more gently folded but highly fractured into large structural blocks and basins; in S China, the limestones have been closely folded (generally NE-SE). Furthermore, though the areas in N China have suffered much neotectonic movement in the later Tertiary and Quaternary, this neotectonic activity was not on the scale of the southern areas particularly those in the proximity of the Tibetan plateau. However, the eastern part of the S China platform has been tectonically more stable.

2. The erosional history in the late Tertiary and early Quaternary in N and S China also forms a contrast. Periglaciation and loess formation in N China have to some extent helped to preserve the old Tertiary palaeokarst. In the south, great erosion seems to have persisted throughout the Pleistocene; evidence of the great mud flow and mass movement deposits indicate the activity of the landscape. Fluvial erosion, especially by the great quantities of allogenic water, produced the relict hills and fenglin in the south, so that today much of the southern areas are made up of residuals. In some areas, they are being further subjected to chemical corrosion – otherwise not much is happening to them. This can be seen particularly well on the southern Guangxi platform near Laibin, south of Liuzhou.

3. To these factors can be added the differences in rainfall between N and S, particularly in the Quaternary and Holocene and modern periods. The imprint of the earlier landscapes are still much more important in N China. It is possible that, as more work is done on the Quaternary of S China, that we shall find more of the landscape of S China is also palaeokarstic. In N China, modern processes have not yet erased the effects of the Tertiary and Quaternary processes; in S China, these modern processes, because of the greater and more intensive and seasonal rainfall, have been much more effective.

9 High Altitude Karst: The High Mountain Karst of West Sichuan

Much of China is mountainous or high plateau, as 65% of the country is over 1000 m. In areas where carbonate rocks are important, high altitude karst develops. These include high mountain or Alpine karst in West Sichuan on the eastern slopes of the Tibetan plateau and the karst on the Tibetan plateau.

9.1 The High Mountain Karst of West Sichuan

The eastern part of the province of Sichuan consists of the famous red sandstone basin surrounded by NE-SW trending hills and mountains associated with the Yangtse platform. As we have seen, many of these hills and mountain ranges developed in limestones and gave rise to long ranges of band fengcong and belted karst (Zhu Xuewen 1981; Barbary et al. 1991)

The Sichuan basin is bounded in the W by the steeply rising N-S or NNW-SSE trending ranges, which form a faulted edge and the transition to the Tibetan plateau. These ranges are frequently over 3000–4000 m, and many are made up of limestones and form a high mountain karst or alpine karst. Little is known of the geomorphology of these mountains; recently, however, two national nature preserves have been set up in the Minshan, an area of mountains N of Chengdu in N Sichuan (Fig. 61). Both of these reserves are in high mountain karst, as are the Jiuzhaigou area and the Huanglong area.

The Minshan form one of the N-S ranges bordering the Tibet-Quinghai plateau which lies to the west (Fig. 61). The highest peak is Xuebaoding, which is 5588 m a.s.l. The Xuebaoding range marks the limit of the rain and snow of the SE summer monsoon, the ranges and plateau to the west being quite dry. Xuebaoding supports a small glacier (2 km long). The average precipitation in the Minshan is about 800 mm and the mean annual temperature is 1.1 °C (maximum is 29 °C and the minimum is −28 °C). As would be expected, there is a great variation in temperature from summer to winter. There is also great diurnal temperature variation; in the summer, temperatures of 28 °C can occur, with temperatures at or near zero at night. The snow line is about 4500 m. Small bushes (cypress, azalea, buddlea etc.) grow to about 3700 m and there are alpine meadows up to the snow line. The high mountain ranges of W and N Sichuan have long been famous for their flora (Sweinfurth and

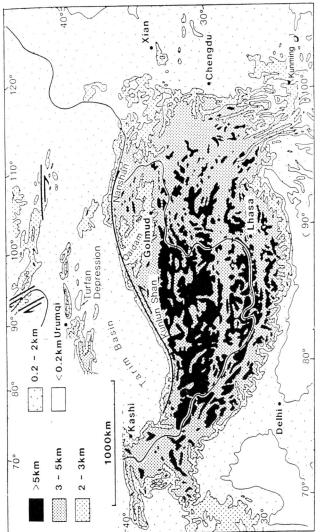

Fig. 61. Simplified topographical map of the Tibetan plateau and Chinese central Asia

Sweinfurther-Marby 1975). The Minshan have large exposures of carbonate rocks of Devonian, Carboniferous and Permian age, which are nearly 5000 m thick.

The ranges are dissected by the deep gorges of rivers that are tributary to the Changjiang (Yangtse) and which flow generally in this part of Sichuan from north to south. They include the Fujiang and the Minjiang. Though biokarstic processes are very important in the limestones in the gorges, landslides and physical weathering are probably the predominant denudation agents on the valley sides. One of the most distinguishing features of the gorges and valleys is the active depositions of calcareous tufa. Water in the high mountain areas above is often quite aggressive with pCO_2 0.23 atm and pH 6; the water discharged from springs in the gorge areas is the discharge zone for this water and the region of sedimentation of the dissolved calcium carbonate (Liu Zaihua et al. 1992). The conspicuous deposition of travertine and tufa in the Minshan valleys is a feature which is not so characteristic of Alpine karst in the European Alps, but is more typical of karst in Bosnia and Herzegovina or Greece and in the Taurus mountains of Turkey – mountainous areas which are not nearly so high as the Minshan and which have a different climatic environment, but similar tectonic environment.

The mountain ranges above 3700 m give rise to pinnacle karst, frost-formed peaks of denudation, similar to those in parts of Tibet. The pinnacles are much smaller than the cones in a fengcong landscape. Also, frost-shattered limestone debris litter their slopes and talus cones and screes accompany the pinnacles. As has been pointed out in connection with the cone karst of Guizhou and in Guangxi, scree formation and frost deposits are relatively rare in the tropical and subtropical parts of southern China. The Minshan pinnacles may be 20–50 m high with broad or narrow bases. Frost-formed rock shelters may occur on steep cliffs and may be about 10 m in diameter and from 5–6 m high (Yuan Daoxian et al. 1991). In zones of intense rock fractures, small caves of about 1 m diameter or less may be found; these result from both frost disintegration and the effects of solution by aggressive snow and melt waters. Frost-formed natural arches can also be found. As in other areas of high mountain karst, superficial solution features are less important, though there are small rillen-karren and solution pits produced by rain and snow; algae, lichens and moss also cause biological solutional weathering of the rock. As a result of the solubility of CO_2, even at these high altitudes, surface and underground water is quite aggresive and the saturation indices of calcite and of dolomite are -1.61 and -3.61. This water infiltrates through the limestones to re-emerge in the valleys. Geochemical work is now being undertaken by different groups in these high mountain areas. Results from the Tibet plateau and the Himalayas done by Nanjing, Manchester and Oxford Universities are given in Chapter 10. Seasonal permafrost is probably present above 3700 m with indications of patterned ground. The mountains in the Minshan resemble the European Dolomites.

Two valleys in the Minshan studied by geomorphologists from the Karst Institute in Guilin and by the Geography Department of Nanjing University will now be discussed in detail.

1. *Jiuzhaigou* (Nine Village valley)

Juizhaigou National Conservation area consists of two south-north draining valleys at an altitude of 3100–2100 m. Limestones from the Devonian to the Triassic in close-set folds (trend NW-SE) occur in the valleys. The valleys lie in the forested zone (mainly coniferous) in the upper parts, but mixed vegetation is found below. The average annual temperature is about 7 °C and the precipitation in the lower parts of the valleys is about 600 mm. Over 40% of the area is forested. This area experienced glaciation in the late Pleistocene and a well-defined moraine occurs in the upper part of the Eastern valley at an altitude of 3100 m. The moraine deposits are calcareous. Tufaceous deposits within the moraine have helped to produce a distinct barrier to the river, forming a lake above.

The main characteristic of the Jiuzhaigou region is the formation of tufa barriers which dam more than 100 lakes (Photo 29). These lakes and barriers occur in both the western and eastern valleys, but are better developed in the western tributary; they continue after the junction of the two tributaries downstream. They stretch over 20 km. Much of the drainage in the Eastern valley is underground (Zhang Jie 1987).

Photo 29. Tufa waterfall in the Juizhaigou valley

Tufa is deposited and extracted from the river waters by two processes; first, by the loss of carbon dioxide and degassing as the water falls over rocks and steep gradients. This is indicated by changes in the pH of lake waters (pH 7.8–8.0) as they pass over the falls (pH 8.3). The second process is by the extraction of $CaCO_3$ from the waters by algae and mosses and also by the trapping of $CaCO_3$ particles by the plant roots. These are also the mechanisms by which the tufa barriers formed in the Plitvice lakes in Croatia, which the barriers and lakes in Juizhaigou resemble quite closely. The tufa is mostly of the soft porous variety, but harder deposits of crystalline travertine also formed. Zhang Jie (1987) studied the different textures of the tufas and travertines. He cited the following different types: (1) crystalline banded texture of travertine laminate; (2) oolitic texture; the ooliths adhere to the surfaces of twigs and aquatic plants; (3) foamed texture found around the mouths of sulphate springs; (4) alga-formed texture; algae contribute to the calcite deposition by binding and trapping and by calcification of the filaments; (5) moss-formed texture, originating when the calcite is formed on mosses. The main algae involved are rheophilous algae (*Diatoma* and *Cymbella* sp.) which are conspicuous in the local algae of Juizhaigou. The plants, mosses and algae also combine to cause surface solution of both the limestones and the tufas, and minor solution features, rillen-karren, rain pits, and honeycomb pits can be seen. Micro-boreholes seen only under SEM and formed by epilithic lichen are also common. As at Plitvice, the effect of the algae is to give the lakes a translucent blue colour; the colours of the Juizhaigou waters add to the beauty of the area.

The eastern valley (Zhezhawa creek) below the glacial moraine at 3100 m consists of long narrow lakes isolated by stretches of dry valleys. The highest lake, Long lake, is over 3000 m. The lakes are connected by underground channels and their water tables fluctuate independently. Screes and cliffs line the lake sides and landslide activity is important. Springs feed the lakes, from both upstream and from the karstic mountains to the E and W, and are associated with tufa deposition. The valley is aligned along the lines of the folding. This valley resembles the *Maligne* valley in the Canadian Rockies (Ford 1979). It is possible that the part below the moraine was glaciated at an earlier period in the Quaternary.

The Western valley (the Rizhe creek) has a permanent stream which connects a fine series of lakes separated by many tufaceous barriers. The uppermost, Swan lake (2900 m a.s.l.) is dammed by an old fossil tufa barrier 30–40 m high and 200–300 m wide. Below Swan lake are a series of lakes with active present-day tufa deposition; the colours caused by the presence of algae are like those in areas of marine coral reef deposition.

A hydrochemical comparison of the waters and lakes of Zhezhawa creek and Rizhe creek is given in the work (Zhang Jie 1987; Table 18).

An important feature in the Rizhe valley is the occurrence of sheet tufa or travertine which forms the so-called Pearl beach; this consists of sheets of hard travertine (crystalline) in the bed of the river and which stretches for

Table 18. Hydrochemistry of the two creeks in Juixhaigou, W Sichuan. (Zhang Jie 1987)

Zhezhawa creek Eastern valley			Rizhe creek Western valley	
HCO_3 mg/l	Long lake	131.78	Arrow Bamboo lake	208.07
	Multicoloured lake	132.64	Panda lake	195.36
	Upper part of river	143.34	Colourful lake	213.85
	Lower part of river	154.50	Mirror lake	201.14
	At confluence of Rizhe	108.66		
Ca mg/l	Lakes as above and waters as above	33.69	Lakes as above	56.69
		34.57		54.23
		36.97		53.82
		39.03		53.82
		32.04		
Mg mg/l	Lakes as above	8.89	Lakes as above	11.47
		8.97		9.47
		10.47		14.46
		11.72		12.71
		4.49		
Total hardness	Lakes as above	6.79	Lakes as above	10.57
		6.90		9.79
		7.60		10.88
		3.40		10.49
		5.53		

some tens of metres. It is quite a different type of calcite deposition from most of the barriers in Juizhaigou which are largely formed of soft amorphous tufa, deposited around moss and plant rootlets. Where the water falls over steep gradients, "beehive"-type tufa deposits are quite common.

The heads of both the eastern and western valleys reach above the coniferous forest line (over 4000 m), and are probably in the seasonal permafrost zone.

The tufa barriers and lakes continue below the confluence of the two valleys. Here, deposition is even more active and the barriers can be over 200 m wide and the lakes up to 2000 m long. The Nuorilang dam is over 15 m high; most deposition takes place where the gradients are steepest. Alkaline-loving vegetation is particularly important in this lower section of the valley; the roots of willow (*salix*) and poplar (*populus*) help to entrap the tufaceous particles and the growth of *phragmites* and other water plants help further the deposition. Where plant growth is vigorous, the ponded water becomes overgrown and the lakes subdivide as below the Shuzheng lakes towards the northern end of the valley. Tufa deposition declines rapidly at the junction of the Juizhaigou river with the major stream.

The age of the tufa is presumably postglacial and Holocene, as suggested by the relationship of the tufa to the morainic deposits at the head of the Juizhaigou river. As far as the writer is aware, there radiocarbon dates on the deposits of Jiuzhaigou suggest 2000–3000 B.P.; the dates for the deposits of Huanglong are given below. A study by Shi Zhenbing (1988) suggests that four main periods of tufa formation can be recognized in Jiuzhaigou from the beginning of the postglacial in this area. Compared with the lakes at Plitvice in Croatia, the tufa barriers and lakes at Juizhaigou are smaller. However, Plitvice lakes are at a much lower altitude – about 400 m. The streams at Plitvice have relatively steep gradients, but are not of alpine proportions. Roglič believed that the occurrence of dolomite was important for the development of the algae and mosses in the Korana river at Plitvice (Roglič 1939). Tufa and travertine deposits occur in other areas of former Jugoslavia and Greece, one of the first on the subject being that by Gregory who called them constructive waterfalls (Gregory 1911).

2. *Huanglong* (Yellow Dragon ravine)

Huanglong is situated south of Jiuzhaigou, also in the Minshan. It is orientated from N to S and is at a height of 3600–3100 m a.s.l. It is situated in an active tectonic zone bordered by the Minjiang fault; 15 earthquakes greater than Richter scale 5 have taken place here since 1949. High dipping (75°) Carboniferous limestones and lower Permian (Maokou) limestones outcrop in the area of Huanglong. As in the Jiuzhaigou valley, there is glacial morainic material at the head, and probable glacial features in the main valley. The travertine and tufa deposits rest upon the glacial forms. The average annual temperature of Huanglong is 1.1 °C and the annual precipitation is about 760 mm. Except for the head of the valley, it is situated in the forest zone.

The most spectacular features of Huanglong are the successions of flowstone (or rimstone) dams and algal pools (Fig. 62 and Photo 30). The flowstone dams are similar to the rimstone pools (gours), which are normally associated with caves (such as Padirac in France and Croesus cave in N Tasmania). Such rimstone pools as at Huanglong are much rarer in surface environments. The pools are of various sizes and depths and are colonized by different varieties of algae, which are of different colours (white, green, blue, yellow etc.). The algae also absorb the light of different wavelengths, with the result that the water in the pools show different colour spectra. The deposited $CaCo_3$ is usually yellow – hence, the name Huanglong (Yellow Dragon). There are over 3000 pools in the valley, from the largest of over 1000 m² in area to those of only 1 m² or so. The pools can be up to 10 m deep. The height of the flowstone dams is also quite variable, the highest being about 7 m high, forming an arc-like dam of 60 m long; the higher barriers occur where the valley

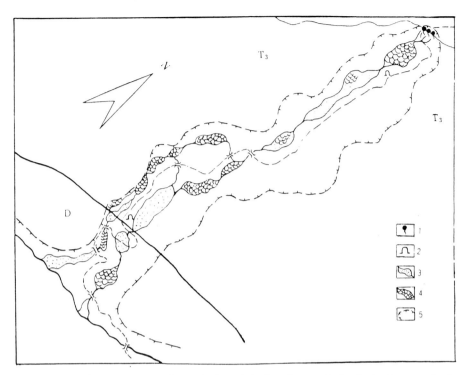

Fig. 62. Sketch map of Huanglong Ravine. *1* Spring; *2* cave; *3* shoal; *4* colour pools; *5* boundary of scenic spot; *T* Triassic rocks; *D* Devonian. The total length of Huanglong Ravine is approximately 5 km

gradients are steeper. There is also a long tufa flow or sinter rapids similar to the Pearl beach deposit in Juizhaigou; the Jinshapudi (Golden Sand beach) in Huanglong is over 2500 m long and up to 170 m wide (Photo 31). Small caves are associated with both the tufa flows and the rimstone dams, which indicate that solution processes are occasionally important in this area of predominant deposition.

The calcite deposited in Huanglong has a different structure and texture from the mainly soft amorphous tufa of Juizhaigou. It is harder and more crystalline in texture and columnar, laminar, and platy in structure. The head of Huanglong is also marked by a glacial moraine, and it is from this moraine that the springs emerge to form the Huanglong river. Zhu Yuanfeng gives details of variations in the water from the springs to the tufa flow (Jinshapudi), which is 3 km downstream (see Table 19; Lui 1991). The spring water also contains 60 tritium (T) units, oxygen–18 is –14.2‰, and deuterium is –99.2‰.

It will be seen that the free CO_2 in the water is very high at the springs; this is partly due to the absorption of CO_2 by snow, but Zhu Yuan-feng also suggests that the high CO_2 in the underground waters may be due to metamorphic processes associated with the neotectonism. This conclusion is confirmed by ^{13}C measurements of the CO_2 (Lui Zaihua et al 1992).

The High Mountain Karst of West Sichuan

Photo 30. Huanglong, tufa dams

Photo 31. Huanglong, tufa terrace

Table 19. Hydrochemistry of waters in part of the Huanglong valley. (Lui Zhaihua 1991)

Location	pH	Total dissolved solids	Ca mg l^{-1}	HCO$_3$ mg l^{-1}	Mg mg l^{-1}	SO$_4$ mg l^{-1}	Free CO$_2$ mg l^{-1}
Springs	6.68	635.57	219.64	742.0	22.9	18.5	165.44
Tufa flow (Jinshapudi)	7.27	377.20	121.60	408.5	18.6	16.6	–

The tufa deposits in Huanglong were dated by ^{14}C, which shows that the tufa grew more or less continuously from the later Pleistocene to the Holocene; the main period of formation was from 8000–2000 B.P. The total thickness of the tufa is generally less than 30 m. Two stone towers and stone tablets were built towards the head of the valley in the Ming Dynasty in 1620 A.D.; these were buried by the tufa up to 1.8–2.0 m in thickness. Based on this observation, the average rate of deposition at this locality was about 5.0–5.4 mm year for the last 270 years. In other parts of the valley, 3–5 mm year have been recorded; at Yingbin pool, the average deposition rate from the 1965–1966 records was 4–5 mm/year. Zhu Dehao in Yuan Daoxian reports values ranging from 0.4–33 mm/year (Yuan Daoxian et al. 1991).

Little is known of the dynamics of the deposition and formation of the rimstone dams. This statement also applies to rimstone dams (or gours) in cave passages. Bretz (1956) said many years ago that "Rimstone growth is a kind of epidemic in some cave streams". Dreybrodt could still say in 1988 that "there are only a few studies on calcite precipitation in surface streams" (1988). Studies, so far, seem to stress the effects of outgassing of CO_2. As at Juizhaigou, the loss of CO_2 over waterfalls and rapids is an important contributor to the deposition process – but all the pools in Huanglong contain algae which may be fundamental to tufa formation. Yet, algae are not usually known to be involved in the formation of rimstone dams in cave passages where they have been described. This theme has been further developed by Lui Zaihua et al. (1992) who state that preliminary results supported by ^{13}C data suggest that the calcite precipitation in Huanglong is entirely controlled by inorganic processes. They measured the evolution of the hydrochemistry with respect to major ions at different locations downstream in Huanglong (Table 20 overleaf). From their data, the measured rates were compared with those calculated by the Plummer et al. rate law (1978); in general, the rates calculated by the PWP-rate law were too high, since Plummer et al. do not consider the existence of a diffusion boundary layer (Liu Zaihua et al. 1992). Hydrodynamic conditions, i.e. flow velocity, influences the value of the diffusion layer and the experiments conducted by Lui Zaihua et al. (1992) were performed under different hydrodynamic regimes, i.e. fast and slow flow conditions. The rates of deposition in the slower flowing pools were lower by a

Table 20. Major chemical parameters from sites 9-1 downstream for both field campaigns in September 1991 and June 1992 in Huanglong. (Lui Zaihua et al. 1992)

Site no.	Field campaign	T_W (°C)	pH	SpC (μS)	Ca^{2+} (mol/l)	Mg^{2+} (mol/l)	HCO_3^- (mol/l)	PCO_2 (atm)	SI calcite	Ca/Mg
9	1991	6.0	6.44	999	$4.93 \cdot 10^{-3}$	$8.93 \cdot 10^{-4}$	$1.22 \cdot 10^{-2}$	0.23	−0.19	5.40
	1992	6.3	6.41	1027	$5.05 \cdot 10^{-3}$	$8.21 \cdot 10^{-4}$	$1.27 \cdot 10^{-2}$	0.26	−0.20	5.96
8	1991	5.1	8.04	870	$4.48 \cdot 10^{-3}$	$8.18 \cdot 10^{-4}$	$1.07 \cdot 10^{-2}$	$5.2 \cdot 10^{-3}$	1.30	5.44
	1992	7.8	7.86	895	$4.37 \cdot 10^{-3}$	$7.31 \cdot 10^{-4}$	$1.08 \cdot 10^{-2}$	$7.9 \cdot 10^{-3}$	1.16	5.76
7	1991	4.3	8.26	588	$3.21 \cdot 10^{-3}$	$6.21 \cdot 10^{-4}$	$6.77 \cdot 10^{-3}$	$2.0 \cdot 10^{-3}$	1.20	5.14
	1992	4.9	8.13	431	$2.00 \cdot 10^{-3}$	$4.15 \cdot 10^{-4}$	$4.89 \cdot 10^{-3}$	$2.0 \cdot 10^{-3}$	0.77	4.80
6	1991	3.6	8.36	461	$2.56 \cdot 10^{-3}$	$5.13 \cdot 10^{-4}$	$5.08 \cdot 10^{-3}$	$1.2 \cdot 10^{-3}$	1.09	4.98
	1992	5.3	8.33	356	$1.63 \cdot 10^{-3}$	$2.57 \cdot 10^{-4}$	$4.36 \cdot 10^{-3}$	$1.1 \cdot 10^{-3}$	0.85	4.58
5	1991	4.0	8.38	497	$2.60 \cdot 10^{-3}$	$6.10 \cdot 10^{-4}$	$5.53 \cdot 10^{-3}$	$1.3 \cdot 10^{-3}$	1.15	4.27
	1992	5.7	8.44	413	$1.82 \cdot 10^{-3}$	$4.34 \cdot 10^{-4}$	$4.95 \cdot 10^{-3}$	$9.8 \cdot 10^{-4}$	1.06	4.19
4	1991	3.3	8.45	482	$2.49 \cdot 10^{-3}$	$6.17 \cdot 10^{-4}$	$5.31 \cdot 10^{-3}$	$1.0 \cdot 10^{-3}$	1.18	4.04
	1992	5.7	8.49	424	$1.90 \cdot 10^{-3}$	$4.34 \cdot 10^{-4}$	$4.80 \cdot 10^{-3}$	$8.5 \cdot 10^{-4}$	1.11	4.39
3	1991	5.4	8.48	445	$2.34 \cdot 10^{-3}$	$6.22 \cdot 10^{-4}$	$4.82 \cdot 10^{-3}$	$8.7 \cdot 10^{-4}$	1.17	3.77
	1992	5.6	8.41	401	$1.66 \cdot 10^{-3}$	$4.65 \cdot 10^{-4}$	$4.32 \cdot 10^{-3}$	$9.2 \cdot 10^{-4}$	0.93	3.56
2	1991	5.3	8.35	414	$2.20 \cdot 10^{-3}$	$6.22 \cdot 10^{-4}$	$4.71 \cdot 10^{-3}$	$1.1 \cdot 10^{-3}$	1.14	3.54
	1992	6.4	8.47	392	$1.56 \cdot 10^{-3}$	$4.72 \cdot 10^{-4}$	$4.72 \cdot 10^{-3}$	$8.8 \cdot 10^{-4}$	1.02	3.33
1	1991	6.3	8.42	397	$2.09 \cdot 10^{-3}$	$6.10 \cdot 10^{-4}$	$4.49 \cdot 10^{-3}$	$9.3 \cdot 10^{-4}$	1.06	3.43
	1992	7.0	8.43	339	$1.65 \cdot 10^{-3}$	$4.94 \cdot 10^{-4}$	$4.31 \cdot 10^{-3}$	$8.8 \cdot 10^{-4}$	1.06	3.43

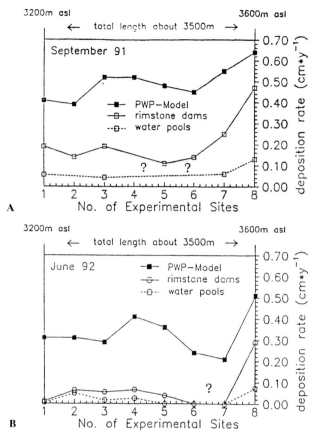

Fig. 63. Calcite deposition rates from experiments in Huanglong valley. (Liu Zaihua et al. 1992) **A** Calcite deposition rates in dependence on various hydrodynamic conditions from field campaign in September 1991. *Solid lines through filled squares*, maximum rates predicted by the PWP rate law, calculated from hydrochemical data in Table 1. *Solid lines through open squares*, measured calcite deposition rates at the rimstone dams (fast flow). *Dashed lines through open squares*, measured clacite deposition rates inside the water pools (slow flow). Lost samples are indicated by ??. **B** Calcite deposition rates in dependence on various hydrodynamic conditions from field campaign in June 1992. *Solid lines through filled squares*, maximum rates predicted by the PWP rate law, calculated from hydrochemical data in Table 1. *Solid lines through open circles*, measured calcite deposition rates at the rimstone dams (fast flow). *Dashed lines through open circles*, measured calcite deposition rates inside the water pools (slow flow). Precipitation rates $< 1 \times 10^{-3}$ cm/year are indicated by?

factor of 3 to 4 compared with those in the faster flowing water on the rimstone dams (Fig. 63).

Zhu Dehao in Yuan Daoxian (1991) makes an interesting comparison of the waters of the Jiuzhaigou and Huanglong valleys. In particular, he shows

Table 21. Temperature, dissolved CO_2 and carbonate saturation indices of karst spring waters in Juizhaigou and Huanglong. (Zhu Dehao in Yuan Daoxian et al. 1991)

Type of karst water		Water temp. (°C)	pH	CO_2 (mg/l)	Saturation index calcite	Saturation index dolomite	log (P_{CO_2})
Spring from aeration zone including surface water	Jiuzai ravine	2.5–13.8	7.79	5.3	0.02	0.73	−1.81
	Huanglong ravine	3.7–7.0	7.70	12.5	−0.4	−1.67	−2.26
Spring from shallow saturation zone		6.2	6.70	139.9	−0.03	−1.02	−1.03
Spring from deep saturation zone		10–22.8	6.49	512.3	0.55	0.03	−0.05

Table 22. Hydrochemical compositions of karst waters in Juizhai and Huanglong valleys

Types of karst water		Content of main ions (mg/l)				Total hardness (German)	Carbonate hardness (German)	TDS mg l⁻¹
		Ca^{2+}	Mg^{2+}	HCO_3^-	O_4^{2-}			
Spring from aeration zone including surface water	Jiuzhai ravine	44.86	11.10	166.90	19.40	165	125.60	1640
	Huanglong ravine	93.70	15.40	348.0	15.30	300	284.80	3020
Spring from shallow saturation zone		201.5	22.9	689.3	18.50	605	589.60	6080
Spring from deep saturation zone		486.4	32.9	1676.8	4.33	1335	1330.00	13890

that the karst springs in both areas are of three types: springs from the aeration zone; springs from the shallow saturation zone; and springs from the deep saturation zone. It will be seen from Table 21 that springs from the deep saturation zone have very high free CO_2 values and also hardness values.

These figures again indicate that geothermal and metamorphic processes influence the deeper waters (Table 22), and increase the CO_2 content. In addition, there is no doubt that cold and snowy mountain waters are able to dissolve appreciable quantities of CO_2 and $CaCO_3$. At 5000 m in Tibet, $CaCO_3$ is as high as 60 mg/l. As these waters descend by steep gradients into much warmer valleys, outgassing of CO_2 is considerable and precipitation takes place; in both Jiuzhaigou and Huanglong, the streams descend from over 5000 to 3000 m in a few kilometres.

Fossil travertine basins similar to those in Huanglong have been described from Antalya by Burger (1990). North of Antalya, Burger shows that "there is a sequence of travertine basins... This sequence proves also that from the top to the foot of the escarpment, water ran from one basin to the next". As in Huanglong "the dimensions of the basins are correlated to the gradient of the slopes along the escarpments. On steep slopes the large basins are of some 100 m in diameter, on gentle slopes, smaller basins have a diameter of some metres." (Burger 1990). Today, the walls of these basins in Antalya have been more or less destroyed, and the interior fillings of the basins have also been eroded. The travertine and tufas associated with these basins have been dated by the uranium/thorium method. The oldest datable travertine is 300 000 ka and is 230 m a.s.l.; the minimum age is 200 ka, which corresponds to the end of stage 9 of the isotopic curve of Shackleton and Opdyke (1973). There are also cones of alluvial tufa in the same area at lower heights with an age of 100 ka, corresponding to stage 5 of Shackleton and Opdyke (1973). Most datings of the tufa and travertine in Antalya correspond to warmer periods of the

Table 23. Karst features of the Minshan peaks and gorges. (Zhu Xuewen, in Yuan Daoxian et al. 1991)

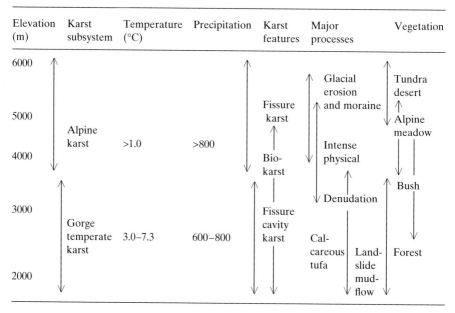

Quaternary (Burger 1990). Unlike the travertines and tufa at Huanglong, those in Antalya are on the spurs and not in the valleys. However, the relationship of the travertine basins to the mountains is similar – the Taurus mountains rising rapidly behind the basins to over 1000 m.

The Chinese geomorphologists make a distinction between the high alpine karst of the mountain areas of the Minshan and the gorges and valleys associated with them. They regard the gorges as *gorge temperate karst* and as a separate subsystem. This distinction would probably not be made in Europe or N America – the two subsystems being treated together. Zhu Xuewen gives a table of the main characteristics of the Minshan peaks and the gorges (Table 23).

The reasons for the extensive deposition of tufa and travertine in these two valleys in Sichuan are probably related to the importance of the neotectonism in this area and an underground environment, whereby the underground water is able to absorb large quantities of CO_2. In similar high altitude areas in the Alps and in the Rockies, tufa and travertine deposits like those at Huanglong and Jiuzhaigou are less conspicuous.

10 The Karst of Tibet and Other Parts of Chinese Central Asia

10.1 Introduction

These areas of mainly high-altitude karst occur in the western half of China, west of the Hengduan-Longmen-Luipan-Helan mountains line (see Sect. 1.1 and Figs. 4 and 61). It is almost half the total area of China. Limestones occur over much of this western tectonic megaregion, particularly in Tibet, where about one-fifth of the terrain is karstic (Fig. 1). All this vast area is arid, with an aridity index of one or more (Yuan Daoxian et al. 1991), and with an annual average rainfall of 100–400 mm and in parts like the Tarim basin (Tali-mu) only 20–25 mm. The average annual temperature is about 2–6 °C outside the basins and 8–10 °C inside them (Yuan Daoxian et al., 1991, p. 110), but the climate is still strongly seasonal. The area includes Xinjiang, Qinghai, Inner Mongolia, Gansu and Tibet and stretches from the Himalayas in the south to Lake Baikal in the north. It is a zone of north-south compressional deformation stretching from the Himalayas to Baikal (Molnar and Deng cited in Dewey et al. 1988). Within this zone of deformation are rigid regions such as the Tarim and Qaidam basins (Fig. 61). The tectonic evolution of the high Tibet plateau was discussed by Dewey et al. (1988). They argue that a thick underthrust of the Indian Shield does not underlie Tibet as often thought, and discuss alternative views. They contend that rapid lithospheric thinning by deformation or stoping would account for the very rapid and recent uplift of the Tibetan plateau, its widespread recent volcanism, its hot springs and E-W extension (Dewey, et al. 1988). "The Indian lithosphere may be thought of as an indenting buttress with a thinner Northern edge generated by Neotethyan Triassic/Jurassic rifting, which collapsed to form the Himalayan Zone of shortening" (Dewey et al. 1988).

Carbonate rocks outcrop widely in bands in Tibet, but north of Tibet they outcrop only sporadically. The limestones occur in many rock formations and systems from the Sinian through the Triassic and Jurassic, and into the Cretaceous and Eocene. Erosional periods gave rise to karst development and buried palaeokarst is widespread. Buried palaeokarst in the Ordovician limestones in the Tarim basin is an important reservoir for natural oil and gas, particularly in limestones with open fissures. The Tertiary and early Pleistocene climates were probably hot and wet in Tibet and in the areas to the north in Xinjiang and Qinghai.

The Kunlun Shan are situated on the northern side of the Qinghai-Tibet plateau and are about 2500 km from W to E (Fig. 61). Karst occurs in the eastern section of the central Kunlun, at an altitude of about 500 m. The carbonate rocks are chiefly lower Permian in age and outcrop for over 3000 km^2. Little geomorphological work has been done in this area, but it has been reconnoitred by Wang Fubao of Nanjing University (pers. comm. 1993). The area is generally arid and frost pinnacle karst seems to be the main present-day landforms. Pinnacles can be up to 40–70 m high and in places are 100 m. Scree deposits are very important. Small solutional karst features are found on the pediments and in valleys. Small present-day glaciers also occur in the Kunlun and where they are associated with the limestones, caves and solution features formed by glacial meltwater are important; they are interesting karstic landforms in an otherwise cold and scree desert. There are cave deposits and speleothems in the caves, which should be of help in reconstructing the palaeoclimates of the Kunlun. The Kunlun is an area of potential study for the palaeoclimatology of this part of central Asia. There are occasional sink holes at the foot of the Kunlun (Yuan Daoxian 1991).

10.2 The General Context of Karst of Tibet

Limestones outcrop widely in Tibet (Xizang). Zhang Dian states that over 230 000 km^2 of carbonate rocks occur in Tibet and Qinghai, out of a total area of 2.5 million km^2 (Zhang Dian 1991). In Tibet proper, there are about 115 000 km^2 of carbonate rocks in an area of 463 000 km^2. The Tibet plateau is about 2000 km from E to W and up to 800 km from N-S. Much of the plateau is over 5000 m a.s.l. and forms the highest continuous tract of land on the earth's surface. The most striking geomorphological feature of Tibet is the evenness of the summit levels and the rolling plateau surface. The surface is by no means level or flat, mountain ranges like the Gangdise Shan and the Tanggula Shan rise up to 7000 m and in the eastern part the rivers have cut deeply into the plateau. The present height of the plateau surface varies from 4500–6000 m a.s.l. Profiles of the plateau from S to N are given in Fig. 64 (Shackleton and Chang 1988). The drainage of the Tibetan plateau can be separated into several systems; the northern part drains into the Tarim and Qaidam basins; the eastern and south-eastern into the Jinsha, Lancang and Nujiang systems and the Yanling Zangbo river (following the Indus-Zangbo structure); the south through the Himalayas as antecedent rivers; and the very large area of internal drainage into the lakes in the western half of the plateau. The main Tibetan highway from Lhasa to Golmud (in Qinghai) basically forms the divide between the western internal drainage area and southern and eastern Tibet which drains into the large rivers of E and SE Asia (Changjiang, Salween, Mekong, Irrawaddy etc.; Fig. 65).

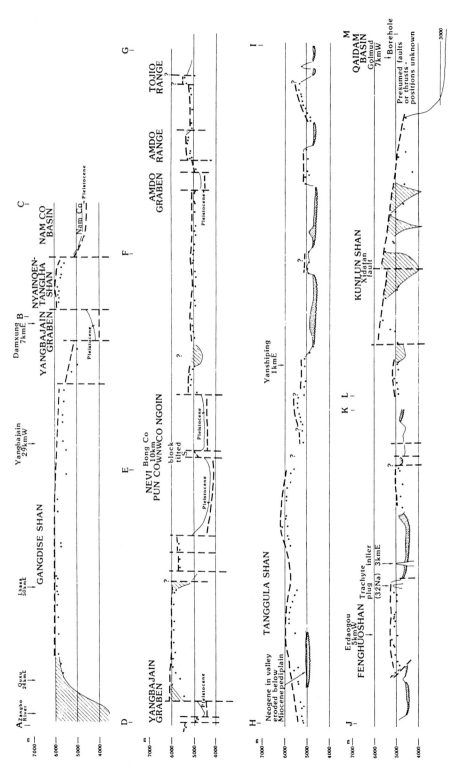

Fig. 64. The Tibetan peneplain. Profiles are based on elevations of peaks (*dots*). Estimated position of pediplain shown by *heavy broken lines*. Post-pediplain faults are shown as vertical since their hade is not known. *Diagonal shading* indicates major post-pediplain erosion. Neogene deposits are *stippled*. Thin lines indicate topographic surface where relevant. (Shackleton and Chang Chengfa 1988)

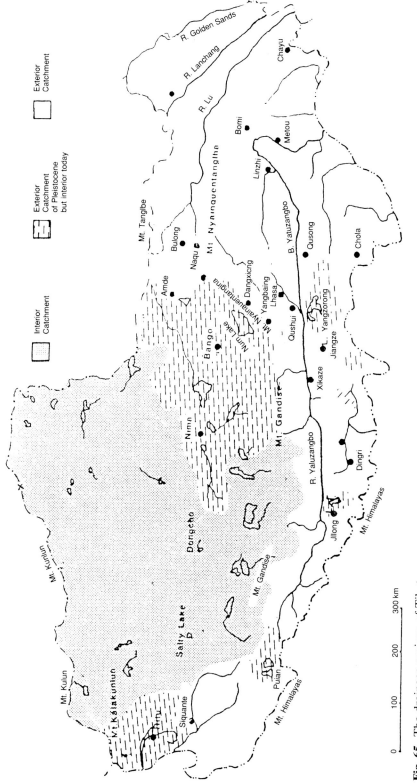

Fig. 65. The drainage regions of Tibet

The widespread Tibetan erosion surface cuts across pre-Neogene, but never Neogene rocks. It is believed to be mid-late Miocene in age, ca. 10 Ma. By the mid-Miocene, the vast area of Tibet had been reduced to a relatively even surface from which only monadnocks (or inselbergs) projected; its elevation was probably about 1000–1500 m a.s.l. in the Miocene under a warm humid savannah climate. By the late Miocene, differential uplift was taking place; although the elevation at that time was more than 1500 m a.s.l., deciduous broad-leaved forests flourished. The Tibet erosion surface or peneplain, therefore, originated under a climatic regime substantially different from that which exists on the plateau today. By the Pliocene, the erosion surface had been severely dislocated by faults and deeply eroded. Pliocene deposits on the Tibet peneplain contain floras which range from tropical forest to subtropical steppe (Lin and Wu 1981). Zhang Zonghu (1991) maintain that the major uplift of the Tibet plateau took place in the Pleistocene which is when the main change in the Tibetan climate is believed to have occurred; the rain bearing monsoonal winds were cut off and, together with the great uplift, the Tibetan plateau was left dry and cold and elevated. Chinese geologists also believe that further uplift took place in the late Pleistocene (Lin and Wu 1981). Apart from the extensive erosion surface, the main feature of Tibetan geomorphology today is the abundance of rock debris caused by frost action. Long slopes of periglacial scree stretch from the mountain peaks to the alluvial fans and debris cones in the broad fault-bounded valleys. Clean rock exposures are rare, except where exposed by down-cutting rivers. Deflation by the strong winds in the arid climate is also important; particularly in the area around the Lhasa river and in winter. Periglacial processes are dominant today throughout almost all areas of Tibet over 4500 m a.s.l. and permafrost occurs in northern Tibet between 4200 and 5000 m altitude. Thermokarstic depressions are abundant in the north – reaching a density north of Amdo of 164 per km^2 but only a few per km^2 in the areas south of Amdo (Zhang Dian 1991).

The area of internal drainage is approximately 624 000 km^2 (about 51% of the total area); land externally drained is about 581 000 km^2. Groundwater contributes more than 22% of the total runoff and meltwater contributes 17%. As would be expected, runoff is highly seasonal, from June-September, over 65% takes place, but from November-April only 11–31%. The period of minimum flow is February with 2% of the total runoff. In the internal drainage area, only a few rivers are perennial and most cease to flow in winter.

The surface deposits blown by the wind form dunes and loess deposits. Dunes are important in the E-W orientated valleys as in the mid-Yarlangzanbo. Loess occurs all over the Tibetan plateau, but thick loess deposits are rare, because they are usually mixed with other deposits. Westerly winds are the most important in dune formation; katabatic winds caused by differential solar heating are important in dune formation in the Lhasa valley (Zhang Dian 1991). Aeolian erosional landforms are not generally important, except in northern Tibet where there are yardangs and ventifacts.

The Tibetan landforms can be divided into three main areas – southern Tibet, including the north slopes of the Himalayas; northern Tibet; and eastern Tibet (Fig. 65). Southern Tibet lies between the Himalayas and the Gandise and Nyainqentangla and includes the high Himalayan peaks and the wide valleys of southern Tibet, including the middle Yanlungzanbo. Northern Tibet lies between the Gandise and the Tanglha, the Kalakunlun (Karakoram) and the Kunlun Shan. It forms the Changtang plateau with an average elevation of 5000 m. The surrounding high mountains have small glaciers; the remainder of the Changtang plateau consists of internally drained basins. The relative relief is of the order of 100–500 m and this area is considered to be part of the Tertiary erosion surface. The northern part is in the permafrost area. Eastern Tibet includes the N-S trending Henduan mountains and is on the sloping edge of the Tibet plateau. The rivers draining into E and SE Asia are deeply cut into this part of the plateau (up to 1500–2000 m). Eastern Tibet receives more rainfall than the other regions of Tibet and in the higher parts, (over 5200 m) modern glaciers exist. The geomnpholopy of western Tibet had been examined by Fort and Dollfus (1989).

As indicated in Chapter 2, there has been much Chinese interest in Tibet; this is particularly so regarding the opinion that the pinnacle karst on the Tibet plateau is palaeokarstic (Chui Zhijiu 1983), and not essentially formed by modern processes active in present-day Tibet. This chapter examines this contention and also owes much to the comprehensive study of the Tibet karst by Zhang Dian (1991).

10.3 Chemical Features of Karst Groundwaters in Tibet

In general, pCO_2 (carbon dioxide pressure) decreases as atmospheric pressure decreases with altitude. This is shown in Fig. 66, which gives measurements of pCO_2 from the Sichuan basin at 1000 m to the Tibetan plateau at over 5000 m. However, in fissures and caves in the rocks, pCO_2 can be above the more general values; measurements of pCO_2 made in soils, caves, sediments and rock fissures at different elevations are given in Fig. 67. These measurements were all made with a standard Dräger meter. The figures for soil CO_2 partial pressures are higher in summer than in winter. The free CO_2 contents of the hot and warm springs that have been examined are much higher and must be due to geochemical processes (Sweeting et al. 1991).

The main radical associated with Tibetan waters is HCO_3 and carbonate waters are dominant in many areas of Tibet. The conductivity of waters in the karst areas arise mainly from the concentrations of $Ca(HCO_3)_2$ and reflect the tendency of $CaCO_3$ to decrease as altitude increases. pH values, however, tend to decrease as altitude *decreases*; this is due to the HCO_3 concentrations which originate from greater biological activity at lower altitudes, as on the slopes of Mt. Zebri in the Dingri area (Himalayan foothills) in southern Tibet (Fig. 68).

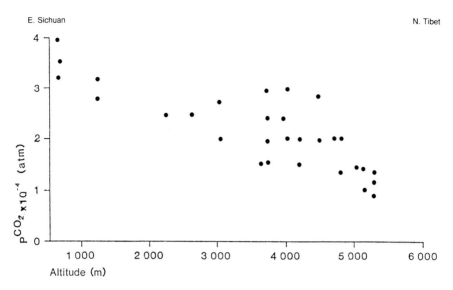

Fig. 66. Atmospheric pCO_2 with changes in elevation from Sichuan to Tibet. (Zhang Dian 1991)

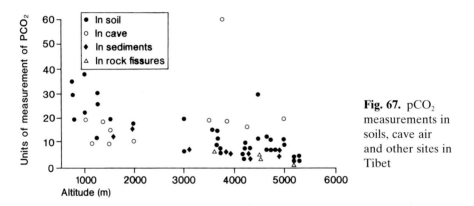

Fig. 67. pCO_2 measurements in soils, cave air and other sites in Tibet

Compared with the carbonate contents of karst waters in other parts of the world, the Tibetan figures are among the lowest (Zhang Dian 1991). High calcium contents in waters in the hot springs are the result of waters from the deep phreatic and geothermal zones. The hot springs have high Ca, Mg, Na and S and often rare elements from igneous rocks. Many of the lake waters are also highly mineralized, partly due to the intense evaporation in the dry climate. All the Tibetan rain and snow samples examined by Zhang Dian had pH values of well over 8.

According to the average pCO_2 of the atmosphere in Tibet, the $CaCO_3$ dissolved in the karstic waters should be less than 60 mg/l, but most localities exceed this level. Zhang Dian explains this (1) by the evaporation of waters in open systems; (2) by external waters from non-carbonate rocks; and (3) by the

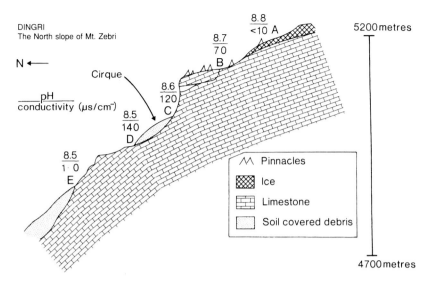

Fig. 68. Chemical changes in snow melt on the northern slopes of Mt. Zebri, southern Tibet. (Zhang Dian 1991)

influence of biological organisms (Zhang Dian 1991). To this should be added the effects of highly mineralized deep groundwaters.

Limestones of Triassic, Jurassic and Cretaceous age outcrop in northern Tibet near Amdo, and in central Tibet near Nam Lake and Lhasa. Limestones of various ages from the Ordovician to the Eocene occur in overthrust masses in southern Tibet, in the Himalayan foothills south of the Yalung Zangbo. Most are fine-grained micritic limestones and are generally thinly bedded. Experiments by Zhang Dian show that there is little difference in their relative solubilities and that petrological differences are not important in differentiating the karst solution processes. Field solution experiments of limestone tablets such as has been carried out in many parts of the world and in other parts of China by Yuan Daoxian (1991), were also conducted in northern and southern Tibet. These indicated that the solution rates at the soil surface and in detrital sediments were higher than the samples exposed only to the atmosphere; Zhang concludes that organic CO_2 and organic acids are, therefore, important in karst dissolution in Tibet. The solubility differences between the major carbonate rock types in Tibet are small; the main factors controlling karst solution are environmental, i.e. climate, geomorphological location and biological activity. Solution rates were, however, smaller in northern Tibet near Amdo, compared with those in Lhasa and other southerly locations (Zhang Dian 1991).

Zhang Dian and also the Tibetan Geological Survey have analyzed some spring groups. Massively bedded and thick Jurassic limestones outcrop on Mt. West near Lhasa (Fig. 69). These limestones are enclosed and intruded by magmatic rocks of the Yanshan and Himalayan periods. Eight springs occur

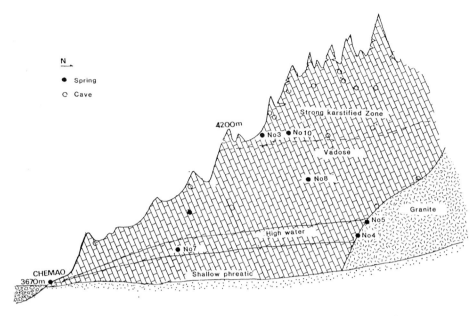

Fig. 69. Distribution of springs on the slopes of Mt. West, near Lhasa. (Zhang Dian 1991)

on Mt. West, the biggest being the Chemao spring. Chemao spring has a discharge of 10–25 l/s in winter and 50–70 l/s in summer; its temperature ranges from 7.9 to 10.1 °C. It is basically a carbonate spring with a mean value of 28.89 mg/l Ca and 3.38 mg/l Mg. These chemical contents of the spring increase with increase in runoff distance and the contact time of the water with the limestones. The reactions of springs to rainfall in the Vadose zone are the fastest. The springs on Mt. West have different discharge hydrographs; springs 3 and 10 have sharp hydrographs, indicating that their aquifers have large cavities and that the water flow is mainly conduit flow; whereas the discharge regime of the bigger Chemao spring has a gentle rise and fall curve, showing that the aquifer structure mainly consists of fissures and pores and that the water flow in it is in the form of diffuse flow. The hydrographs also reflect the influences of rain and snow; after snow the peaks are smooth and symmetrical; after rain the hydrograph peak is sharp and asymmetric. The vadose springs (springs Nos. 3, 8 and 10, see Fig. 69) have a greater temperature variation, whereas the Chemao (shallow phreatic) spring shows a smaller variation (Zhang Dian 1991).

The Jiala spring in the Jiala village, 20 km from Lhasa is in an aquifer of Carboniferous and Permian limestones. The spring is fed by meltwater, and the spring discharge increases after every period of rise of surface temperature, but the discharge regime peaks do not respond to the times of precipitation. The Tibetan Geological Survey (1975, unpubl) has also investigated the Queshang springs, in the Triassic Chaqupu group and situated 60 km from Lhasa. These springs consist of cold and hot springs which belong to different drainage systems. The water flow of the cold spring comes from the aquifer near the surface and its mineral contents are much lower than those of the hot spring. The temperature of the cold spring is about 3 °C, while that of the hot spring is 48–49 °C. The flow of the hot spring passes through adjacent magmatic rocks. As in other hot and warm spring areas of Tibet, the Queshang hot spring outlet is associated with much tufa deposition (Photo 32).

Such studies indicate the existence of different aquifer structures in the Tibetan limestones. Aquifers exist with united water levels, which can be divided into various hydrogeological zones; other aquifer structures have independent conduits where hydrogeological zones are difficult to distinguish. The existence of big cavities and passages in the vadose zone on Mt. West, in Lhasa, implies a former strongly karstified zone now above 5100 m. Circulation of deep underground water is slow, with long lag times; high pressures and temperatures in the deep phreatic zone give rise to spring water with high solute concentrations, high temperatures and more stable behaviour (Zhang Dian 1991).

10.4 Surface Karst Features

As already indicated, the limestones in Tibet, though there are many types, are usually thinner bedded than those in the platform regions of S China. Lithological and structural differences must always be borne in mind when making comparisons between the forms in Tibet and those in S China. The

Photo 32. Hot spring and karst resurgence in tufa north of Amdo, Tibet

results given here are the first comparisons to be drawn. The most important karst geomorphological features of the Tibetan karst areas are the limestone pinnacles which develop on the harder beds (Photo 33). These are associated with scree aprons separated by water or snow gullies. These pinnacles were regarded by Chui as being similar to the fenglin and fengcong of S China (Chui Zhijiu 1979). They are particularly well developed in the Lhasa area (Photo 33). Zhang Dian has made a morphological study of these pinnacles. His measurements of the parameters of pinnacles is shown in Fig. 70. Individual

Photo 33. Periglacial pinnacle karst, near Lhasa

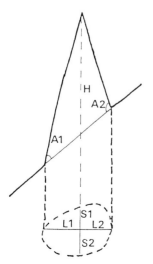

Fig. 70. Methods for measuring geometrical parameters on Tibetan pinnacles. *H* Height; *L* longer axis; *S* shorter axis *A* angle of slope. (After Zhang Dian 1991)

pinnacles were measured in the Dingri (59), the Lhasa (89) and the Amdo (56) areas, both in the field and from field photographs. The symmetry of the pinnacles (P) was determined from formula $P = L1/L2 \times S_1/S_2$ and the values of P divided into four groups (1) symmetric ($P = 1 - 1.5$); (2) subsymmetric (1.5 – 2.0); (3) asymmetric (2 – 3.0); and (4) very asymmetric ($P > 3$). Zhang also applied this formula to the karst towers (fenglin) and the cones (fengcong) of S China. The results are shown in Fig. 71; the pinnacles in the Tibetan karsts are asymmetric to very asymmetric, while the towers and cones in S China are symmetric to subsymmetric (Zhang Dian 1991).

The plan form of the pinnacles represented by the ratio of long to short axes (l/s) was also determined for the three areas: Dingri, Lhasa and Amdo (Fig. 72). The distribution of angles of pinnacle slopes shows that the angles for the same rocks and areas are quite irregular and scattered in distribution, whereas the distribution for the karst towers and cones in S China (Shuicheng) fall within a narrow range. The mean relative heights of pinnacles in the three Tibetan karst areas (15–25 m) are much less than those of the cones and towers in the tropical and subtropical areas (50–300 m) (Fig. 72). There is thus a difference in form between the Tibet pinnacles and the tropical and subtropical karst towers and cones, though not enough to conclude that the pinnacles are *not* the remains of tropical karst. The strike of the strata controls the direction of the long axes of the pinnacles in the Zebri mountain area, but in the Lhasa and Amdo areas, joints control the directions of the long axes.

All the pinnacles in Tibet are located on the tops of the hills or mountains or on their slopes. The relative height of the pinnacles decreases with decrease in slope angle; debris from their weathering accumulates below, by the downslope movement of debris. Such accumulation of debris is much rarer in the formation of cones (fengcong) and towers (fenglin). Zhang Dian also points out that the cones (fengcong) are associated with a polygonal karst hydrological network; the Tibetan pinnacles are associated with "normal" gully systems. Their irregular and asymmetric plan and the large differences in height, angle and diameter/height ratios together with their relationships to the strike of the rocks and joints suggest that the pinnacle formation is caused by physical weathering along weak planes and joints in the strata.

Zhang Dian has analyzed the pinnacle patterns in the Tibetan karst areas and suggests that there are two main types (Fig. 73).

1. *Lhasa type*: These resemble semi-arid and arid slope profiles, with four components (upper convexity, free face, straight uniform slope and basal concave slope). Gullies caused by ephemeral water or debris flows are incised into the cliff face and increase the height of the pinnacles; the steeper the slopes the deeper the gullies.
2. *Amdo type*: The slope profiles of the limestones of Mt. North in the Amdo area are periglacial; the pinnacles are like "tors" with altiplanation terraces between them. Because of the extreme frost weathering and weak fluvial activity, the height of the pinnacles is less than in the Lhasa area. The

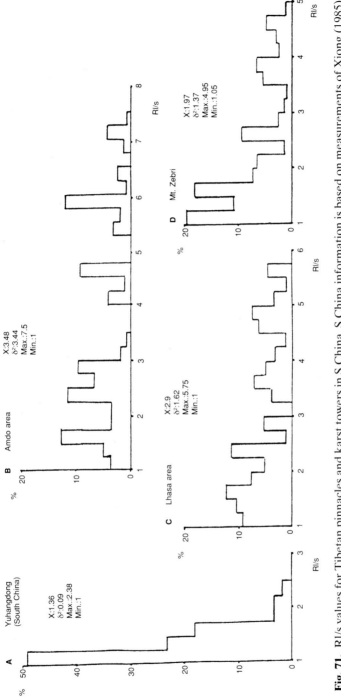

Fig. 71. Rl/s values for Tibetan pinnacles and karst towers in S China. S China information is based on measurements of Xiong (1985)

Surface Karst Features

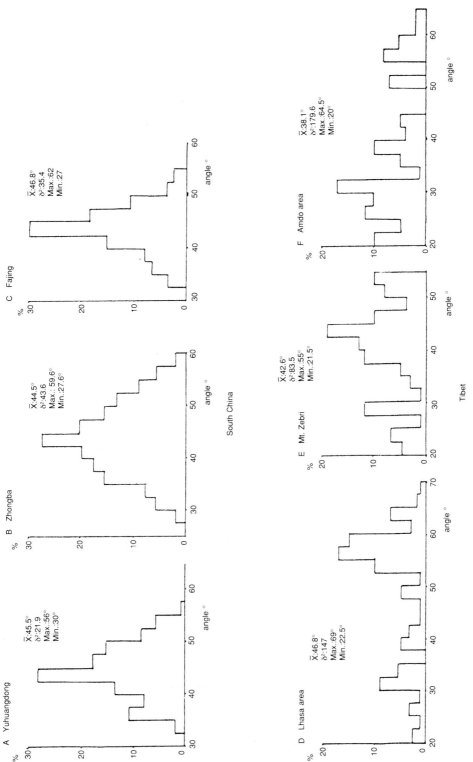

Fig. 72. Slope angles of Tibetan pinnacles and karst towers in S China. S China information is based on measurements of Xiong (1985)

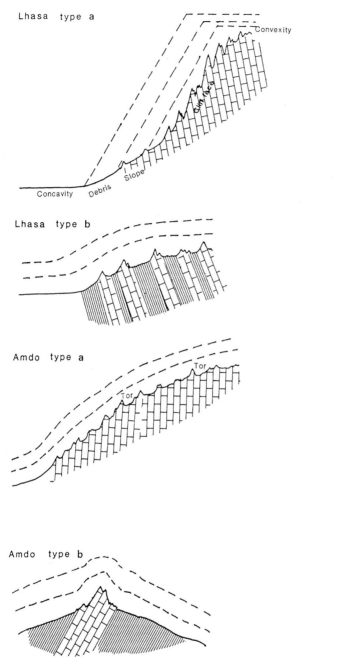

Fig. 73. Evolution models of Tibetan limestone pinnacles. (Zhang Dian 1991)

altiplanation terraces are covered by polygonal soils; solifluction processes lead to microrelief forms and solifluction lobes on hill slopes of 15°–20°. Solifluction and stone flow are more lateral than vertical and this fact widens the distance between the pinnacles (Zhang Dian 1991; Photo 34).

The pinnacles on Mt. Zebri are intermediate between the Lhasa and Amdo types. On the upper part of the mountain, the pinnacles resemble those of Amdo, but lower down, the pinnacles have been dissected and enlarged by meltwater and debris flows and are more like those in the Lhasa area.

Despite the relatively weak solutional activity of the karst waters on the Tibetan plateau, and the fierce effects of frost weathering upon the limestones, solutional forms are found on the Tibetan limestones. These are classified by Zhang Dian, and their importance in the different areas given.

Rain pits occur throughout Tibet in groups on horizontal and inclined bare limestone surfaces. The deepest (up to 9.8 mm) occur in the Lhasa area, possibly because of the higher precipitation in the Lhasa valley (485 mm/year). In some areas, a thin tufa film (0.2–3 mm thick) covers the base of the rain pits giving a smoothed outline. Frost action destroys the original solutional forms in high mountain areas. Rain pits occur more in areas where the precipitation is in the form of rain rather than snow.

Rillen-karren (solution flutes) are found on many limestones in Tibet. They develop on inclined outcrops of fine-grained pure limestones and marbles in all areas, on slopes from 10° to 74°. The usually sharp crests between the runnels are often destroyed by frost action, except those in the Lhasa

Photo 34. Karst near Amdo, Tibet

valley. In the Lhasa valley also, a clear relationship between the length of the rillen-karren and the degree of slope of the rock can be observed. Elsewhere in Tibet, it is probable that present-day frost action prevents the free development of rillen-karren.

Solution pans are basin-shaped closed depressions on horizontal limestone surfaces that are larger than rain pits. They may be open or covered. Covered pans are covered by soil and vegetation, such as fungi and mosses; they are clearly enlarged by the solution of water with high biological CO_2 content.

Grikes or *kluftkarren* are formed by solution along joints or fissures in the limestones in all areas. *Wall solution runnels* develop on the vertical or inclined walls of limestone outcrops. *Solution troughs* are formed at the contact zones between polygonal soils and the limestone outcrops.

10.5 Caves and Cave Sediments

Both the author and Zhang Dian have spent much time looking for caves in Tibet. All Tibetan caves are small, mostly under 20 m long and the longest so far discovered in the areas we examined is 100 m (Photo 35). The highest cave found so far is at the top of Mt. Zebri at 5600 m. The caves are usually 1–3 m, but occasionally may be up to 18 m in diameter. Zhang Dian concludes that the

Photo 35. Shallow caves in the cliffs at Nam lake

caves are relict, "not developing at present, but formed in some past time" (1991). Because of the extreme tectonic movements that the Tibetan limestones have suffered, the limestones are cut by many joints, particularly sheared joints. Sheared joints are more permeable than the bedding planes and favour water flow. Joint development is related to the age of the strata, the oldest beds have most multidirectional joints and the most complicated direction patterns of the cave passages. This is well illustrated in the more complicated caves in the Triassic limestones of the Amdo area in northern Tibet. Hydraulic gradients are also important in determining cave directions; in Mt. Zebri cave passages are aligned in the direction of the mountain slopes (Zhang Dian 1991).

As already indicated, all caves so far discovered in Tibet are "dead" caves, with no present-day water flow. The caves are related to palaeowater flows (Fig. 74). Zhang Dian classified the Tibetan caves in terms of water flow, i.e. phreatic, vadose and compound (when two or more hydrogeological types of water flow are involved). Caves in the Lhasa and Amdo areas have a large proportion of phreatic type caves, implying that these limestones were below the water table for a long period. Caves in the Lhasa area experienced several stages from phreatic to vadose and back again as base levels fluctuated; Big Buddha cave is an example. Near Dingri in the Himalayan foothills, vadose caves are predominant on Mt. Zebri, implying that cave development occurred after the formation of Caves on Mt. Zebri are most often related to the hydraulic gradient; they have steep lower parts and straight passages. Many caves on Mt. Zebri are situated between 5000–5200 m a.s.l. just within the palaeo-snow line and close to the glacial ice limit; meltwater is believed by Zhang Dian to be important in the cave formation in this area. In the Mt. West area of Lhasa, the caves in the higher parts are phreatic and largely joint-dependent; below 4100 m the caves in Mt. West are vadose in type. Inflow caves on Mt. West are substantially fewer than the outflow caves; Zhang Dian interprets this as possibly due to small snow or glacial inlets on Mt. West, amalgamating lower down the mountain into bigger outflow caves.

Sediments, both on the surface and in caves, are associated with the Tibetan karst landforms. Some of these are still forming today (for example the travertines and tufas around the hot springs), while others, such as the stalactitite material in many of the caves, formed in the Quaternary or Holocene. Travertine is quite often deposited around the outlets of hot and warm springs; CO_2 is dissolved at depth under high pressures, but is diffused into the atmosphere as the springs emerge at the surface. The travertine is often layered, as at the Quesang Hot spring, and at the warm spring, north of Amdo.

Both travertine and tufa are associated with calcareous deposition by both surface and underground waters. As a result of the dry climate, rapid evaporation after rainfall gives rise to tufa films on rock outcrops, up to a few millimetres thick. Thicker laminar tufa up to 20 cm thick occurs in some of the more arid areas of southern Tibet. Laminar tufa may form where other mineral salts in addition to calcite are involved; it can also be formed as a

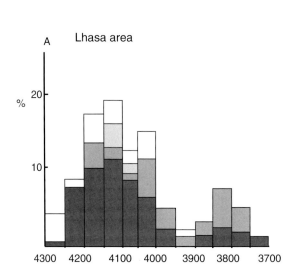
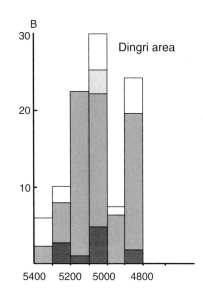
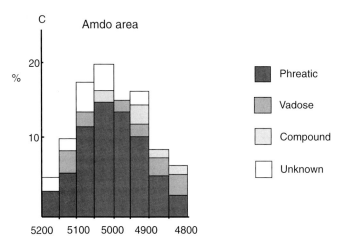

Fig. 74. Distribution of different cave types in three karst areas of Tibet. (Zhang Dian 1991)

subglacial precipitate, as on Mt. Zebri (Ford et al. 1970, 1983). Concretionary tufa occurs in the Amdo area. The evaporation of percolating surface water causes deposition of tufa between limestone blocks on the hillslopes; the tufa occurs in the fissures between limestone blocks and cements them to the limestone outcrops.

In the caves, the sediments are both chemical (stalactites, flowstone etc.) and clastic (clays, silts, gravels etc.). There are very few chemical deposits to be found in the Tibetan caves, and most of these are not forming today. Small stalactites have been found in the Tibetan caves. Flowstone, formed by water flowing over the rock in a thin film, has been found in caves in Lhasa and Amdo. The thickest flowstone deposit so far discovered is in the Big Buddha cave near Lhasa – up to 25 cm thick. Flowstone is often interbedded with clastic layers. Cave coral occurs frequently in the Tibetan caves; this is a capillary deposit related to evaporation. Cave coral forms "knobbly" deposits often on cave walls. It is sometimes laminated and Zhang believes the differences in the laminations may indicate climatic changes from arid to wetter conditions.

Clastic cave sediments may be derived from the internal block breakdown of the limestone rock walls and roofs and from external sources (fluvial, aeolian, or lacustrine). All sizes of clastic cave sediments are found in the Tibetan caves, from clays and silts to gravels and boulder debris. Aeolian and fluvial-glacial sediments are usually allochthonous, while the breakdown sediments and filtrate sediments are normally autochthonous. Aeolian sediments are widespread and are particularly important in the Lhasa area. Fluvial-glacial sediments are important in the Dingri and Mt. Zebri area near the Himalayas. Breakdown sediments caused by mechanical weathering are widespread and their particle size varies from sand to boulder size; they are particularly important at cave entrances. Alluvial sediments in the Tibet caves are normally thin, though thicker layers exist in the caves near Lhasa possibly caused by periods of temporary flooding.

These sediments (both chemical and clastic) have been examined by Zhang Dian (1991) and Bull et al. (1990) to see what light can be thrown upon the processes which caused them and under what type of environment they were formed. Stromatolitic and brecciated structures are closely related to evaporations and frost weathering and often occur in recently formed deposits. But laminated structures in caves are the product of water-rich conditions. The caves are now dry; therefore, the water-rich environment which produced the laminated structures in caves must have existed in the past. Zhang also attempted to produce sequences of cave sediments (Fig. 75).

Grain sizes and shapes reflect weathering characteristics. As is the case for surface sediments, fine sediments in caves are produced by chemical weathering in a warm humid climate; cold, humid climates are associated with mainly coarse clastic materials. The size range of the alluvial cave sediments in the Tibetan caves suggest very slow flow speeds (from 3 to 6 cm/s), indicating water flows associated with chemical deposition. In general, most Tibetan cave samples are poorly sorted, partly resulting from the addition of aeolian and breakdown debris. Looking at the mean size and sorting in the Tibetan sediments, there are two clusters of points; one is the coarse particles with high sorting values and the other is fine particles with low sorting values. Most of the cave alluvial sediments are subangular, implying that the samples were derived from nearby sources; this result is likely as all Tibetan caves examined so far are short.

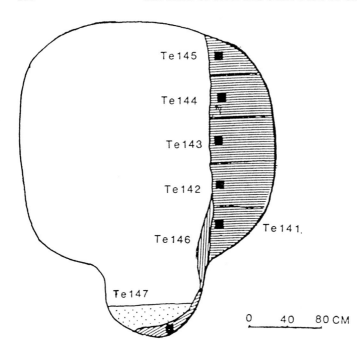

Fig. 75. Cross section of No. 5 cave indicating nature of sedimentary sequence. The *numbers* refer to specimen numbers. (Zhang Dian 1991)

The surface texture of quartz grains from the cave sediments have been examined by Bull et al. (1990) and Zhang Dian (1991). While many of the grains studied are fresh and lack significant environmental modification, samples collected from old caves and depression deposits in the Lhasa area have features which suggest a high energy chemical environment. The evidence indicates that there was at least a warm and wet period in Tibet which led to strong chemical and mechanical features on the surface of the quartz grains.

Zhang Dian also examined the main iron minerals in the Tibetan sediments; these are goethite and haematite. Many of the samples from caves have these iron minerals. Generally, there is a tendency for haematite to develop in soils of warm, well-watered regions and for goethite to form elsewhere. Zhang Dian divided the mineral composition of iron-rich samples in Tibet into five groups: *Group A* (goethite and calcite with little or no quartz). In the Big Buddha cave, near Lhasa, goethite and haematite are deposited with calcite. It is possible that these were deposited by water with a low pH in a warm and wet climate; *Group B* (goethite with haematite, kaolinite, and quartz). Zhang Dian regards these deposits as most likely to be related to laterite deposits and transported by water flow and deposited in caves. They were deposited under a climate warmer and wetter than that of the present day; *Group C* (goethite with original minerals). These are considered as sediments formed in drier and warmer early Pleistocene climates; *Group D* (a mixture of Group A and some original silicate minerals and secondary clay minerals). These samples contain chlorite and are connected with flowstones. They indicate some chemical weathering in a climate that "was not dry and cold"; and *Group E* (a mixture of Group B and some original silicate minerals). These samples were found in karst depressions and in cave sediments. Their structures indicate a mixing during deposition.

The silicate composition of many Tibetan karst sediments are related to the nearby silicate-rich rocks, usually granites. In the Tibetan sediment samples so far examined, montmorillonite appears only in one sample, whereas kaolinite is present in many samples. Those containing kaolinite occur in cave alluvial sediments and in depression sediments.

All these analyses, together with the presence of some flowstone and stalactites, indicate that there were periods when the climate in Tibet was warmer and wetter than it is today. In the sample examined, there were quite fresh and unaltered samples and there were chemically weathered samples. These different types occur on the Tibet plateau and indicate that the environmental change was very fast; this is in accordance with the theory that the Tibet plateau rose to its present height very rapidly (Han Tonglin 1991; Zhang Zhonghu 1991).

Speleothems are rare in the caves of Tibet and most of those examined have been discovered in the Lhasa area. The stalactites and flowstones from the Tibetan caves have been dated by Uranium series dating. The Tibetan speleothems are difficult to deal with as many are quite contaminated and the dates may be subject to age errors (D.C. Ford, pers. comm.). Zhang Dian (1991) gives the results from the analyses by Ford. The main conclusions are as follows:

1. Most of the cave flowstones are more than 350 ka and may be older than 1.25 million years. Many of these old speleothems are in caves in the Lhasa area, which also contains the oldest caves.
2. Some speleothems developed between 20–350 ka, suggesting that there were some warmer and wetter periods in the mid-late Pleistocene.
3. The tufas are generally 5000–6000 B.P. but tufa at Dingri gave a date of about 100 ka and was formed in a dry and cold period.
4. The dating of a cave coral sample in the cave at 5250 m a.s.l. on Mt. Zebri gave an age of 25 ± 2 ka. The lack of flowstone in the Mt. Zebri caves implies that they are younger than those in the Lhasa area.
5. Only one cave containing flowstone was found in the Amdo area, in northern Tibet, at 4700 m a.s.l. The flowstone has been dated as forming between 25–175 ka.

The surface clastic sediments, except for those in the depressions (of which there are only one or two), have all the characteristics of sediments produced in cold and dry climates. Chemical weathering is very weak, and physical weathering is the major agent. The formation of these sediments correlates with present-day climatic conditions; older surface clastic sediments have not been found. In the depressions, the sediments are a mixture of those with intense chemical corrosion and those formed under present-day conditions.

In the caves the flowstones and stalactites were formed in warmer and wetter climates than prevail today. The cave alluvial sediments contain kaolinite and much goethite and have chemical and water flow features on the quartz grains, and were formed between the flowstone layers. These alluvial sediments were transported into the caves under warmer and wetter conditions than those of today. Cave breakdown and breccias formed after the flowstones; these sediments have fresh, unweathered quartz, angular grains and poor sorting. They cover the cave floors and the older speleothems.

10.6 Some Conclusions on the Tibetan Karst

The analysis of the Tibetan sediments thus suggests the following climatic variations:
1. A Tertiary tropical and subtropical period. The evidence is derived from lateritic deposit fragments from the Tertiary erosion surface and now found in the cave and depression sediments.
2. An early and mid-Pleistocene warmer and wetter phase compared with today. Many of the dated speleothems, and cave alluvial and depression sediments are of this period. Some sediments associated with this group indicate a cold and dry climate, suggesting some cold and dry subperiods.
3. The cold and arid period from the late mid-Pleistocene to the present. The limit between the warmer and wetter climate and the cold and arid period is regarded as the late mid-Pleistocene, 220 ka ago. Since this time, extreme physical weathering has prevailed, though there have been some moist and cool substages, proved by dated cave "corals", soil layers, flowstone and peat.

From his detailed work in the Tibetan karst areas, Zhang Dian comes to the following conclusions. In the *Lhasa area*, the karst evolution can be divided into three major stages. The first stage is that of phreatic cave formation in the Pliocene; the second stage is of vadose cave and speleothem formation which occurred in the early and middle Pleistocene; and in the third stage the development of caves and speleothems ceases and only rain pits, rillen-karren and some small biokarst features formed from the late Pleistocene to the present. There have been slight fluctuations in climate since the late Pleistocene such as the wetter periods, 8000–3000 B.P., and also the drier periods when sands accumulated in the caves and on the surface.

In the *Dingri area*, which is located at 4000–5000 m on the northern slopes of the Himalayas, the karst developed in Eocene limestones. Zhang believes that the karst in the Dingri area began in the mid-Pleistocene and was brought about by a higher precipitation and meltwater. The caves are vadose and fluvial-glacial origin. There are no flowstones or stalactites in the high mountain caves of the Dingri area, as there are in the Lhasa caves of lower altitude; the higher elevation of Mt. Zebri gave rise to a colder climate and flowstone did not form. Since the mid-Pleistocene, only microcaves and karren have formed in the Dingri area. The relatively young development of karst in Dingri is partly caused by the tectonics produced by the proximity of this area to the Main Central Thrust of the Himalayas. As a result of the pressure, limestone beds are broken and convoluted and often only recently exposed. Comments on the lack of cave development in other areas of the Himalayas have been made by Waltham (1991).

In *northern Tibet* the karst landforms are mostly on the Tibet erosion surface at 4500–5000 m. In the Amdo area, as in other parts of northern Tibet, karst pinnacles occur on the upper parts of the hills while the lower parts are covered by aprons of frost debris. The small caves are not later in age than the end of the *Pliocene*; in the Quaternary erosion was mainly under periglacial conditions. Solution forms developing under soil polygons are the only solutional karst features occurring in northern Tibet at the present. The formation of caves in Zaxido island (Nam lake) are related to wave erosion and are late mid-Pleistocene.

No karst landforms dating from the early and middle Tertiary have thus been found by the writer or by Zhang in the areas so far examined, though the climate in the Tertiary was suitable for karst development. One can only conclude that any karst landforms which did occur in these early periods were destroyed by the intensity of the physical weathering processes in the Quaternary and Holocene; from the late mid-Pleistocene, and the period of rapid tectonic change, extreme physical weathering and frost degradation prevailed. The periods in the Holocene of rather warmer and wetter phases only allowed the formation of small solutional features such as rain pits and rillen-karren and small biokarst forms. The present climate in Tibet is regarded by Wang and Fan (1987) as being a cold and dry phase, during which time the aeolian sands accumulated in the caves and on the slopes of the hills. The significance of the tors and tafoni in the Lhasa valley has been examined by Osmaston (1992).

There are still many areas of Tibet to be explored for karst landforms, which may change or modify the views expressed in this chapter. In particular in eastern Tibet, Yuan Daoxian (1991, p. 77) reports many caves on the western bank of the Jinsha river, and which he says are "several hundred metres" long. There are also large caves in West Tibet along the Shiquan river (near the Pakistan border). These caves have large chambers "big enough to hold several hundred people"; they are well decorated with speleothems and have been used for storage. Such karst landforms are clearly very different from those discovered and discussed by the writer and Zhang Dian (1991). Karstic features have been reported from western Tibet by Fort and Dollfuss, (1989). As in other aspects of Tibetan geomorphology and geology, much more systematic study is needed, in both the field areas and in the laboratory study of the sediments.

10.7 The Salt Karst of the Qaidam Depression

North of Tibet are the faulted mountains and basins of Qinghai and Xinjiang where the aridity index is generally over 4 (Yuan Daoxian et al. 1991). Salt karst can be found here and an area of salt karst occurs in the Qaidam depression, in Qinghai. One part has been studied by Guan Yuhua and Xu (1981) and Guan Yuhua and Song Linhua (1984) in the Qarhan salt lake region of the Qaidam depression. The lakes are situated in the central part of the Qaidam basin in Qinghai province, at an altitude of 2677 m and above, so it is also high-altitude karst. The climate is dry continental, with average annual temperature of 2.5 to 5.1 °C; the average annual precipitation is 28.1 mm, but the annual evaporation is of the order of 3250 mm. The Qarhan Salt lake is an old lake flat with about nine smaller lakes now being preserved; they consist of brine water with a high concentration of salt – over 300 g/l. According to Chen Kezao and Bowler, the lakes are not residuals of palaeo-Qarhan lakes, but have been newly formed since about 9000 B.P.

The features of the salt karst were described by Guan Yuhua and Song Linhua (1984). They include dissolution pores which are up to 5 m deep and from 0.5–3 cm in diameter and are caused by meteoric water. There may be from 4 to 287 pores m^2 beneath the salt crust. The surface of the rock salt is covered by a dissolution crust; the sides of fractured salt may be bent to form knife-like edges, caused by the recrystallization of large salt crystals. Beneath the crust, small salt stalactites may develop. Salt teeth are produced by rainwater dissolution and wind erosion. Rainwater dissolution can also produce small pits; conical pits with diameters from 1–5 cm and depths of 3–5 cm occur and flatter pits 10–50 cm in diameter and 5–20 cm. deep are also found. Shallow karst valleys occur along the northern boundary of the salt belt; they may be some kilometres long. In these valleys, mineral groundwater or shallow confined water rises along the salt lake edge – dissolving the salt. Confined gas is also associated with the confined water and the gas pressure may break through the salt crust to form gas-exploding holes. The mouths of these holes are higher than the surface of the salt crust; their diameter may be 2–10 cm and their depth 20–50 cm. Many of the caves are formed by the combination of dissolution and gas explosions. Around the lake shores are small terraces of salt caused by wave erosion. Guan Yuhua and Song Lin Hua also recognize different types of caves; these include tube-shaped, funnel-shaped and fissure types; none of these are large and reach only a few metres.

A horizontal zonation of the karst is recognized. The zones are: (1) solution valley zone; (2) shallow cave zone; (3) deep cave zone; and (4) solutional pit and pore zone. The salt cave evolution may involve four stages: (a) dissolution by descending water; (b) solutional enlargement; (c) equilibrium of dissolution and crystallization; (d) crystallizing infill (Fig. 76). The formation of these karst features is dependent upon the hydrochemistry. This depends upon the recharge of the salt-karst brine water with lateral water of low total dissolved solids (TDS), which will corrode the rocks. According to measure-

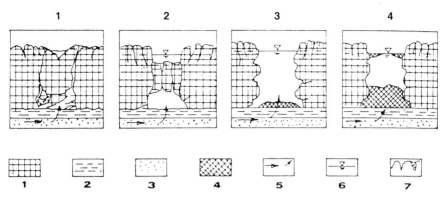

Fig. 76 Schematic diagram of formation and development stages of salt karst caves. 1. old salt — 2. mild clay — 3. silt and fine sand — 4. new salt — 5. confined flow — 6. phreatic water level — 7. solutional fissures — (1) dissolution and descending — (2) solutional enlargement — (3) equilibrium of dissolution and crystalization — (4) crystalizing salt to fill the cave.

ments, the salt karst process can only occur when the specific gravity of the brine is lower than 1.2 and the TDS are less than 310 g/l. When the salt is dissolved, the specific gravity increases, TDS gets higher and reaches saturation; solution then ceases and precipitation may occur. Yuan Daoxian et al. (1991, pp. 111–112) report that this process changes with the seasons (pp. 111/112). The solution rate in zone 3 (see above) has been calculated at 0.6–1.5 cm/year, and the cubic solutional rate is 0.016–0.040 m³/year (Guan and Song 1984).

According to Chen Kezao and Bowler ^{14}C dating in the western and central parts of the Qarhan Salt lake shows that the salt layer at the bottom of the lake was formed about 24 000 B.P. The salt layer in the upper part was formed ca. 16 000 B.P. and deposition of the salt ceased around 9000 B.P. and the lake dried out about then. Later, when the climate became more humid, some new lakes appeared along the southern verge of the dry salt flat. Small corrosion pools developed, together with caves, pits and salt karst springs. The salt karst seen today is believed to have formed within the past 9000 years (Chen Kezao and Bowler 1988).

10.8 Conclusions on the Karst in Tibet and Chinese Central Asia

The huge area considered in this chapter is still relatively unknown from the geomorphological and palaeoclimatic point of view. Much remains to be discovered, particularly relating to the extent to which the karst features were formed by palaeokarstic processes or by the active frost and periglacial processes of today. Because of the arid climate, some sediments formed in a more humid and warmer period in the late Quaternary and Holocene have been preserved, however, further studies along these lines are needed. It is agreed by most geomorphologists that the Tibet plateau did not support a large ice cap in the Quaternary (Shi Yafeng 1992), but this view is still disputed (Kuhle 1991). Until such major questions are resolved, the origin of karst landforms in this part of China will continue to be uncertain. The author believes that the major karst landforms on the Tibetan plateau were formed by frost processes in a cold and arid climate such as exists today; the frost pinnacles are of a completely different origin from the peak-forest landforms of tropical and subtropical South China. However, we know little about the effects of other processes (such as wind); we know even less about the controls which the limestone lithologies, the fracturing of the rocks, and the neotectonics have exerted upon the development of the karst in these areas of China.

11 Karst Hydrogeology and Chemical Characteristics of the Karst Water

11.1 General Comments and Behaviour of Karst Water

Reference has been made several times in the discussions of the karst geomorphology to the springs and underground waters of the karst areas. This chapter gives a summary review of the karst hydrogeology and the characteristics of the karst waters, particularly in relation to the geomorphology. Both this and the concluding chapter review the contributions of Chinese karst to world karst geomorphology, and its position in general world karst studies.

Karst groundwater in China totals about 203.97 billion (10^9) m^3/year, making it about one-quarter of the total groundwater in the country. However, this is unevenly distributed in space and time. Karst water is controlled by the framework of the carbonate rocks and is stored by and flows through many thousands of karst hydrological systems. A karst hydrological system may be defined as a system with definite karstic boundaries containing water and air, which circulate through the karst under dynamic equilibrium. The karstic boundaries are determined by the stratigraphy of the soluble rocks, in relation to the non-soluble rocks, the geological structure and the relief.

A simple spring or underground stream may be a karst hydrological system – with precipitation recharge as the input and its emergence at the spring as output. There are thousands of karst hydrological systems in China. In different parts of China, the karst hydrological systems differ according to the local geological conditions, climate and geomorphology of the area. It is generally asserted that karstification, i.e. the amount of solution undergone by the rocks, is much more intensive in South China than in North China. As we have seen, South China is noted for its numerous underground stream and cave systems, whereas in North China there are very few underground streams as such, but it has some very large springs.

The movement of karst water in the conduit medium may be (1) *longitudinal conduit flow*, i.e. the direction of the underground flow coincides with the trend of the conduit. The nature of the water movement is similar to the flow of the compressive conduit and surface canal. The movement of longitudinal conduit water is in accordance with the hydraulics of conduit and canal; or (2) *lateral conduit flow* when the flow direction is at right angles to the trend of the conduit and has little effect on the movement of conduit flow. The valleys are the locations of the conduits when the water table is at the low stage (Fig. 77; Song Linhua 1981).

Karst water flows in isolated and semi-isolated situations, but also in a unified groundwater. If uplift takes place, unified groundwater flow converts to separate water flows; if subsidence occurs, isolated and separate water flows may be converted into unified groundwater. A classification of major hydrodynamic water conditions is given by Lu Yaoru et al. (1972).

The geological structures are responsible for the development of solution fissure systems, solution pipe systems, and solution pipe networks. Karst groundwater may be divided into four types: fissure flow, pipe flow, vein flow and network flow. As the development of karst weakens with depth, the characteristics of the groundwater change. The distribution of karst groundwater exhibits a zonal feature in vertical section. Generally, it may be divided from the top downwards into the cavity fissure zone, cavity flow zone, and the deep-solution-fissure zone.

The dynamic behaviour of karst water changes as it is recharged. There is a reciprocal transformation of surface and underground water. Groundwater enrichment is usually found where there are many geological structures and in tension fissure zones; it also collects near local base levels of denudation. Groundwater is unconfined in plain and plateau areas and hillslopes, but is confined in the buried karst basins. The runoff conditions and type of flow vary

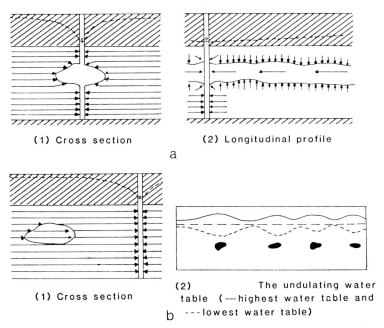

Fig. 77. The two main types of karst water movement in conduit medium. **a** Longitudinal conduit flow; **b** lateral conduit flow. (Song Linhua 1981)

with time. From recharge through discharge areas, the burial depth, hydraulic gradient, flow rate and fluctuations of ground water often show a regular change from the smaller or lower to the bigger or higher and back again.

The different geological and precipitation environments act to make the karst groundwater systems function. The input information, the precipitation, along with the geological structures and lithologies, directly determine the presence and the properties of the systems. The hydrogeology and the geochemistry of the waters are, therefore, a function of the type of karst. Broadly, as in the division of karst types, the Qinling line (34° N lat.) separates two types of groundwater systems. The characteristics of the karst water-bearing media in North and South China are shown in Fig. 78, taken from Cui et al. (Cui Guangzhong 1985).

11.2 Contrasts Between Karst Water in North and South China

In North China, the karst is located in the limestones on the Sino-Korean platform. In this area, since the Palaeozoic era, there have not been great horizontal crustal movements, but there have been eperiogenic uplifts and subsidences of crustal blocks. The folds are gentle and only large faults or fractures affect the carbonates – the middle Proterozoic-Cambro-Ordovician groups. Karstic denudation affected these rocks between the late Ordovician-early Carboniferous and where gypsum beds are present (as in the Majiagou) this has given rise to palaeokarstic "collapse columns" (see Sect. 8.3). The water-bearing medium of the soluble Palaeozoic rock is compact and hard, and a huge corroded fissure network system was formed on the stable platform in the successive carbonate rocks. Only occasionally near the major spring

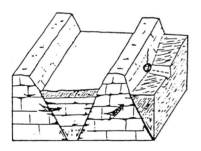
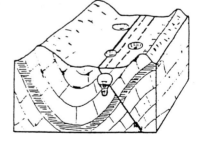

Fissure medium system
in structural block

N. China

Cave-conduit system in
fold and fault zone

S. China

Fig. 78. Pattern showing formation of typical karst water-bearing medium in North (*left*) and South (*right*) China. (After Cui et al. 1985)

outlets is conduit flow found. The karst water systems are broad and large scale.

South China includes the Yangtse platform and the South China geosynclinal fold area. The great thicknesses of carbonates which were deposited upon the Yangtse platform Sinian-upper Ordovician (in the SW) and mid-Devonian to middle Triassic means there are multilayer carbonate deposits with clastic rock partings. Since the Mesozoic, there has been frequent crustal movements, giving rise to tight folds and intensive fracturing. The intensity of the crustal movement increased towards the west. Many closed anticlines and synclines of relatively small areas were formed. The water-bearing media are more heterogeneous than in the north, and form complex systems, including corroded fissures, but also cave-conduit systems. Subsurface flow often moves in underground streams, which are very common in S China. It is calculated that there are over 2800 underground streams (in caves) in S China, but only 8% of them have a length of more than 10 km. The karst water systems are of small and medium scale, resting on the active platform with multilayer fractured and folded carbonate rocks. The total length of the underground streams in S China is nearly 14000 km with a total discharge of 1842 m^3/s (Yuan Daoxian et al. 1991; Fig. 79).

Thus, these differences in the karst water-bearing media give rise to the differences in structure and function of the karst groundwater systems in these two areas. An exception to these general statements is the area of NE China in Liaoning, near Shenyang where the stratigraphy resembles that of N China but where the structures resemble those of S China. In this area there are conduit flows, underground streams, large caves and small-scale karst water basins.

North China is characterized by large karst springs. There are over 150 big springs with discharges of more than 0.1 m^3/s. Their total discharge is more than 200 m^3/s. Over 60 have a mean discharge of more than 1 m^3/s. A map of the large springs in N China has already been given in Fig. 50 (after Yuan Daoxian et al. 1991). A great deal of research has been done by Chinese hydrogeologists into the catchment areas, variations in flow and geological conditions of these large springs; this is because of their importance to agriculture, domestic water supplies, industry and tourism. Boundaries of the catchment of most systems are known and the springs have been monitored for many years – sometimes for hundreds of years.

Because of the tectonics of the N China platform, the Cambrian and Ordovician rocks are gently dipping and are broken up into self-contained fault blocks. The karst hydrogeological systems cover large areas – up to many thousands of square kilometres. Closed depressions are rare but a feature of the landscape is the dry valleys. In addition to direct infiltration into the rock, the loss of surface runoff in the dry valleys is an important part of the input into karst springs. Yuan Daoxian quotes as an example, from Baiyangshu to Luanlui, a distance of 4 km on the Taohe dry valley, the loss of surface water is 0.44 m^3/s; this water becomes part of the discharge of the Niangziguan karst spring, the biggest in N China (Yuan Daoxian 1988). Also, the loss of the Zhuozhang river at Xinangqiao, north of Changzhi city, Shanxi province is 6.15 m³/s. The water lost from the surface valleys, infiltrates, but does not form distinct underground streams, at least not today, under the present conditions of a rather semi-arid or subhumid climate.

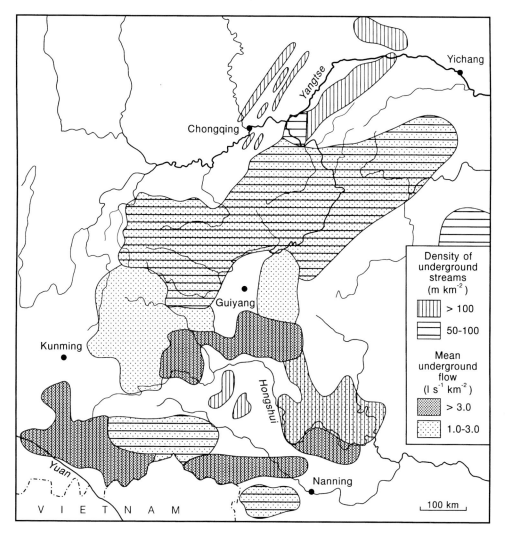

Fig. 79. The density of underground streams and models of underground flow in South China karst. (redrawn by the Cartographic Unit, Department of Geography, University of Manchester after Yuan Daoxian et al., 1991)

The underground flow regime is stable and flow is diffuse. The hydrographs show a stable flow with the amplitude of fluctuation of no more than twice the average height. Moreover, peak discharges usually show several months or even several years' time lag with respect to the relevant rainfall events. For instance, these points are illustrated by the Lonzhichi spring at Linfen (Shanxi); the discharge ranges from 5.4 to 6.7 m^3/s and its hydrograph usually shows a 6-month lag to rainfall (Yuan Daoxian et al. 1991). The time lag of the Nyangziguan spring (Shanxi) is 7 years and that of the Baotu spring (Jinan, Shandong) is 3 years. In studying them, it is important, therefore, to have rainfall records for as long as possible. The input signals are well modulated, because of the large recharge area and the diffuse nature of the flow in the aquifer. The karst systems act as vast underground reservoirs. The flow is generally homogeneous and obeys Darcyian laws. In general, therefore, the transmissivity of the limestones is high, the storage of water very great and the springs large – a great contrast to the conditions in S China. Possibilities of pollution are, however, very great and difficult to eradicate (Yuan Daoxian and Cai Guihong 1987).

In *South China*, underground streams are very numerous. In an area of 300 000 km^2 in the provinces of Guangxi, Guizhou, Sichuan, Hubei and Hunan, there are at least 2836 underground streams. The minimum discharge of these totals 1482 m^3/s and the total length is 13 919 km. A map of the density of underground streams and their modules of flow in S China is given in Fig. 76. Because of the complicated tectonic structure, and because of intercalations of non-soluble rocks, the size of the karst hydrologic systems is variable – from a 1000 km^2 to just a few km^2. The relief affects the distribution – for instance the densities of underground streams are greater on the north and south sides of the Guizhou plateau, where the plateau is deeply dissected by the gorges draining north to the Changjiang system and south to the Hongshui-he in the upper reaches of the Zhujiang (Pearl) river. Here, the hydrodynamic conditions are favourable for the development of caves and underground streams with steep water-table gradients. The longest underground stream so far known in South China is in Disu, Duan county, Guangxi (Fig. 46). This is on the north side of the Hongshui gorge and is developed on the strike of a northwest-trending anticline formed of Devonian to Carboniferous limestones and limited on both flanks by Middle Carboniferous dolomites. The Disu system has a total underground course of 241.1 km, a 57.2-km-long trunk stream and several branches. Its catchment area is 1004 km^2 and its mean discharge is 38 m^3/s but its fluctuations are considerable, from 4 to nearly 550 m^3/s (Yuan Daoxian 1988).

Closed depressions are innumerable in the South China karst, together with collapsed cavern "skylights" and ponors (sink holes or swallets). These are the typical ways of input for precipitation into the karst hydrological systems. Rainfall infiltration rate by means of enclosed depressions is high – as there is now little surface drainage. Infiltration rates may be as high as 70–90%.

In the Disu system there are more than 200 karst windows as inlets to the system. The total catchment is 1000 km^2 and the mean infiltration rate varies from 50 to 80%, with an average of 66%. The mean annual rainfall is 1700 mm with 80% between April and September. The low water flow is 4 m^3/s and the measured maximum flow of the underground river system is 390 m^3/s which is 100 times the low water discharge. The groundwater level at flood time may rise 35 m in its lower reaches and up to 100 m in its upper reaches. The Disu river flows on the surface from April to October, but is underground for the rest of the year. The hydraulic gradient of the underground river is similar to that of the surface river – about 5–12% in the upper reaches and 0.5% in the middle and lower reaches. The flow from swallow hole to underground conduit is quite rapid with a generally quick response of outflow to rainfall; the regulation of the karst system is, therefore, poor. Much research into the behaviour of the system from input to output has been done and the processes of water flow from swallow hole to spring is far from simple. It will be seen that the great discharges of such underground river systems, i.e. up to 390 m^3/s, are agents of great geological erosion and are important in the formation of the big cave systems which occur near the water table. In Yunnan, some underground rivers have a gradient of over 20%, falling 1690 m in 37 km for example.

Most karst springs in South China consist of a combination of conduit flow in underground passages and fissure flow; the conduit flow is usually fed from the swallow holes, but it is regulated by fissure flow. Compared with North China these systems in the south reflect rapidly the storm and intense rainfall events during the monsoon season. The hydrographs show a good correlation with the seasonal variation of precipitation, but also have a very short time lag to the storm. The Penshuidong underground stream in Yunnan has a discharge in the rainy season of 2–6 m^3/s but in the dry season it is only 0.5–1 m^3/s. The time lag of its flood peak to the storm is usually less than 5 days. As a result of the well-developed underground conduits, the storm pulses are not adjusted or regulated within the system. In some cases in South China, springs have an amplitude of discharge of up to over 1000 times (Yuan Daoxian 1988).

Cui Guangzhong. (1985) made a study of two springs in North and South China to illustrate these points – they are the Heilongdong spring in Hebei and the Luota spring group situated at the junction of the four provinces, Guizhou, Sichuan, Hubei and Hunan.

1. In N China the *Heilongdong karst* groundwater system consists of four spring groups that contain 60 spring points. The area of the spring catchment is 2400 km^2 and the main karstified rock is middle Ordovician limestone. The catchment is located on the eastern part of the Shanxi platform; the strike of the strata is NNE and there is a series of strike faults with a high angle which cut across the area. The western barrier of the spring is Precambrian metamorphic rocks and quartzose sandstone; the eastern barrier is the NNE regional fault, the Ordovician limestone making contact with the Permian coal strata. The north and south boundaries consist of impervious rocks. The catchment

area of the spring is, thus, a closed karst groundwater system characterized by a hydrogeological unit in a tectonic block in North China.

Data from 400 boreholes show that the karst underground water system is mainly corroded fissures; the largest cave revealed is only 4.5 m high and is a large corroded fissure. The spring group issues from visible fractures, 0.05 to 0.7 m wide which strike NE to ENE. There are two intensively karstified zones, which strike N-S. They converge towards the outlet of the spring group. The concentrated seepage flow of the karst groundwater means that subsurface flow exists in the intensively corroded zone; there is also a concentrated seepage flow zone.

Rapid seepage flow occurs in the intensively karstified fissure zone, with high transmissivity, which is 40 times more than in the seepage zone. The hydraulic gradient is low – only 0.04–0.36%. The response of the water-level in the concentrated seepage flow zone is more rapid than in the seepage flow zone. The annual fluctuation of water level is 3.4 m in the concentrated seepage zone and 1.8 m in the seepage zone. The main features of the karst flow in the Heilongdong spring area are summarized in Table 24.

Table 24. Features of karst flow zones in the Heilongdong spring area, N China. (Cui Guangzhong 1985)

Geological setting	Shallow fractures in uplifted blocks near the faulted belt	Fissure system in the structural block
	Concentrated seepage flow	Seepage flow
No. of corroded cave fissures found in the bore hole (mean value)	2.2–3.4	0–0.57
Linear rate of karst in water-bearing part (%)	4.2–12.6	0–3.3
Transmissibility (m^3/day)	436 000–84 000	2000–30 000
Hydrochemical type	HC$_3$-CaHCO$_3$-CaMg	HCO$_3$, SO$_4$-Ca Cl.SO$_4$, Na. Ca.
Total dissolved solids	0.4–1.0	1.0–6.0
Hydraulic gradient (%)	0.04–0.36	0.6–1.0
Lag of peak between rainfall and hydrograph in 1979 (days)	55–95	60–104
Annual fluctuation of groundwater level in 1979 (m)	2.26–3.4	1.86

2. The *Luota subterranean* spring system is in S China. In this group, there are 51 underground streams with a total length of 85 km in an area of 106 km². The maximum flow rate in the wet season is 185 m³/s, but it is only 0.2 m³/s in the dry season. It is located in the E Guizhou fold zone of the Yangtse paraplatform; the karst water is in the Permo-Triassic carbonate rocks and its distribution is controlled by fractures. The impermeable basement is of Siluro-Devonian sandy shale. The karst water system exists in the many layers of carbonate rocks and is controlled by the structural fractures. The system is complex with many subsystems. The trunks of the subsurface streams developed along NNE-trending main faults – the branches along tensional faults trending NNW. The underground streams are cavern conduit systems and each subsurface flow, separated by an aquifuge, forms a subsystem.

The hydraulic gradient is quite steep; the maximum gradient is 13% and the maximum velocity can be over 4300 m/day. The response hydrographs of the discharge of subsurface streams to the rainfall pulses have a sharp, large peak and a relatively short time to peak and a rapid recession. In general, the lag of the hydrograph peak is only 1–2 days – and under rapid conditions, it may be only 2 h. The storage and regulation of the system are poor, for instance, in some parts of the Luota system with an unstable discharge, the volume of discharge of rapid flow may be 89.7% of the total water volume (Table 25).

These contrasting characteristics are further illustrated in Table 26 from Cui (1985), giving the difference in the karst groundwater systems between the northern and southern parts of China.

Table 25. Characteristics of the main subsurface streams in the Luota area (S. China). (Grangzhong 1985)

	Main trunk stream (Wuyandong)	Bangbangdong
Hydraulic gradient (%)	26.2	62
Velocity (m/day)	750–2777	754–1092
Discharge (min.) 1/s (max.)	200 157700	39 32770
Coefficient of variation (Q_{max}/Q_{min})	788	840
Lag of hydrograph peak (day)	0.5	2
Rate of rapid discharge in volume (%)	22.9	20.5

Table 26. Karst groundwater systems in North and South China. (Cui Guangzhong 1985)

Structure of system	North Monolayer karstified flow zone	South Monolayer – multiple aquifer karstified flow zone or subterranean stream
Area of system (km^2)	n × 100 – n × 1000	n – n × 100
Input of rainfall mm/year	400–600	900–1600
Minimum discharge of system during dry season (m^3/s)	0.3–10	0.05–1.4
Coefficient of variation Q_{max}/Q_{min}	1.24–5.89	10–1000
Amount of groundwater run/off during dry season (1 s^{-1} km^{-2})	3–6	3–8
Lag of hydrograph peak (day)	n × 10	0.08 – n × 10

The statistical data show that though the area of karst groundwater systems in South China is smaller than in the north, the coefficient of variation of discharge is two orders of magnitude higher than in the north. Although the precipitation of North China is less than that in South China, the discharge of a system in the dry season is larger and more stable. Thus, the geological environments are fundamental in the functioning of the karst groundwater systems. However, in South China, with high precipitation in the mountain areas, the effects of erosion and corrosion by underground streams can change the medium of the karst groundwater system, and produce the feedback to the input of the rainfall. The steep gradients of the surface and underground rivers in S China, particularly in Guizhou, have encouraged the building of power stations, as in the Maotiao river (Yang Mingde et al. 1988). In a study of the karst water resources of China, in relation to their exploitation and harnessing, Lu Yaoru (1987b) estimated that about 17–18% of the total surface runoff in S China and about 18.9% of the total surface runoff in N China is karstic. He also gave the gradients of various karstic rivers in China (see Table 27).

A diagram of the exploitation of the karst water resources in China is given in Fig. 80.

11.3 Karst Water Heterogeneity

The carbonate rocks of China, particularly the pre-Triassic which includes most of the limestone, were compacted, and have a very low primary porosity (generally less than 1%), and the crystalline dolomites may have a primary

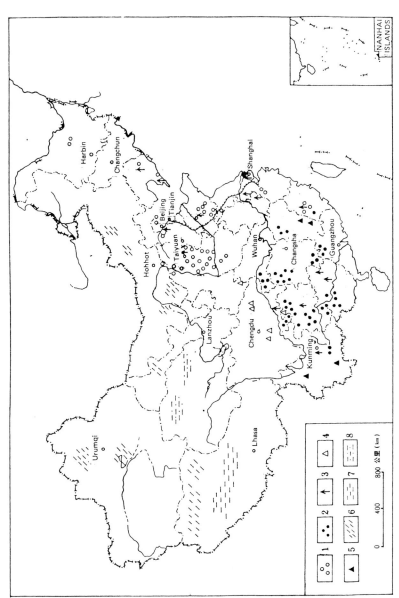

Fig. 80. Exploitation of karst water in China. *1* Typical regions for exploiting karst water resources by wells, tunnels and guiding karst springs. *2* Typical regions for exploiting karst water resources by constructing underground dams and guiding karst springs. *3* Important cities exploiting karst water resources of water supply. *4* Typical region exploiting subsurface salt water and brine. *5* Typical region with rock salts. *6* Typical region with salt water-brackish water lakes for extracting salts and other minerals. *7* Compound karst region enriched in lithium and boron in salt lakes. *8* Compound karst region enriched in potassium, magnesium and boron in salt lakes. (Lu Yaoru 1987b)

Table 27. Gradients of various karstic rivers in China. (Lu Yaoru 1987b)

River	Length (km)	Gradient (‰)
Changjiang (Three Gorges-San Xia)	193	0.2
Wujiang (Guizhou)	1050	1.65
Maotiao He	180	3.05
Dadu He	1062	3.9
Jinshajiang (Sichuan)	1545	1.4
Hongshui He	659	0.3
Lancangjiang (part in China)(Yunnan)	2153	2.0

porosity of about 5%. Karst water, therefore, flows in a variety of karst features in the limestone, rather than circulating in intergranular or intercrystalline pore spaces. Karst water is thus not evenly distributed in the rock, even in the porous limestones of the north. This unevenness of distribution is called karst water heterogeneity. Heterogeneity is, however, more a feature of the South China karst systems. Heterogeneity is most easily expressed by the percentage of high-yield wells in an area compared with the total number of explored wells. Wells sunk into karstified strata may yield hundreds of tons of water per hour, but less than a few metres away the yield may be less than 1 ton. Near Wulixu, near Guilin, 43 boreholes were drilled in the peak forest plain underlain by well-karstified Lower Carboniferous limestones; the boreholes were situated in an area 3 km long and 2 km wide. However, only 14 of them gave enough yield from exploitation and hence the percentage of well completion is 33% which is a moderately hetereogeneous aquifer. In areas of extreme heterogeneity, the well completion rate may be less than 10%.

Heterogeneity of karst water is also reflected in the anisotropic transmissivity and very irregular nature of the cone of depression of a pumped well. A very anisotropic example is shown in an exploration site near Liuzhou, Guangxi (Fig. 81). The aquifer is in Middle Carboniferous dolomite and when pumped the drawdown of the cone of depression stretched 1400 m to the NW, but only 200–300 m to the NE. Heterogeneity varies, depending upon the degree of integration and connection of the karst features in a karstified rock mass. In the Zaoqing-Fofu syncline, Wumin, Guangxi, eight wells were drilled into Middle and Upper Carboniferous limestone. They were 90–120 m deep and drilled in an area 5 × 2 km. The percentage of well completion for this area was 88% (seven out of eight) and the aquifer was less heterogeneous. Heterogeneity can be studied macroscopically (regionally) or microscopically (in hand specimens); but from the practical point of view for hydrogeological exploration, four types of heterogeneity are recognized by Yuan Daoxian (Table 28). These are based on the percentage of well completion and the ratio of axes of the elliptical cone of depression (Yuan Daoxian 1985a–1988).

Heterogeneity is related to the size and integration of the karst features. It is often contended that there is a sequence from the isolated conduit to an integrated network, i.e. a heterogeneous aquifer will evolve into a homogeneous one, as karstification proceeds. However, this is not borne out by results from China. In the south, intensive karstification has been going on, at least, from the beginning of the Tertiary (and possibly earlier), yet the large caves and underground streams have a very heterogeneous distribution. In North China, where karstification is considered to be less intensive, there are many karst aquifers that are hydrologically more or less homogeneous. Hence, heterogeneity and homogeneity depend upon geological circumstances and the manner as well as intensity of karstification. Under homogeneous circumstances, water will accumulate; in heterogeneous circumstances, there will ususally be rapid drainage.

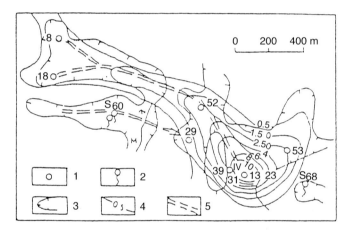

Fig. 81. Irregular cone of depression around No. 31 pumping borehole, Liuzhou, Guangxi. *1* borehole; *2* karst spring; *3* doline; *4* isopiestic line during pumping; *5* presumed underground conduit

Table 28. Well completion and classification of degree of heterogeneity. (Yuan Daoxian 1985a)

Type	Well completion (%)	Axial ratio of cone of depression
Individual cave	5	1
Extremely heterogeneous	20	5
Heterogeneous	20–50	2–5
Relatively homogeneous	80	1–2

11.4 Epikarstic Water

Williams (1983) and Bakalowitz and Mangin (1980) called attention to the epikarstic (or subcutaneous) zone and its importance in karst hydrology. An experimental site set up near Guilin gave some interesting data on the epikarstic zone. The output spring (automatically monitored) is regulated in at least three ways: by the slope soil cover around the closed depressions (dolines); by the epikarst; and by the aeration zone beneath the bottom of the closed depression. Almost every patch of soil on the doline slope has a perched spring, though the discharge is small. However, quite a few such springs may flow the whole year-round and survive the dry season with discharges of 0.1 l/s; the carbonate hardness is about 160 ppm during storms but rises to about 280 ppm in the dry season.

Despite the rough nature of the limestone surface and its many open fissures, the rainfall and slope runoff do not drop directly to the underground stream. The drilling data show that there is an intensively karstified zone which is limited to a depth of about 10 m; this karstified zone is the epikarst. Below the epikarst, the limestone is much less permeable and less corroded. After storms, therefore, infiltrated water flows down the doline slope into the lower part of the epikarst zone. Epikarst springs have a discharge of 30 l/s and flow out of the bottom of the dolines, well above the normal spring outlet. Water from the epikarst springs often drains into swallow holes at the lower end of the dolines and thus contributes to recharge the karst water system. Some of these epikarstic springs may flow the whole year and in the dry season, their discharge may be about one-half of the total output of the system. Their hardness in the dry season is usually about 190–200 ppm $CaCo_3$. These experiments tell us much about the regulatory function of the epikarst zone to the hydrologic system (Yuan Daoxian and Drogue 1988).

The regulation effect of the aeration zone below the bottom of the doline is reflected in the fluctuations of the water-table. Some boreholes in the Guilin experimental site reached the underground passage below the level of the main outlet spring. The water table in this passage has an amplitude of fluctuation of about 30 m; the hydrograph shows a rapid rise after a storm, but a slow recession, indicating that both the underground passage and the fissure network play a regulating role in the system.

A further illustration of the controlling effect of the type of karst upon the hydrology is again shown in the Guilin area. Fenglin (peak forest) and fengcong (peak cluster) exist side by side in the Guilin district. The peak forest plain developed in this area where the phreatic water table is near or at the surface; the plain is also well watered by allogenic water from the Lijiang. Lateral corrosion is predominant and the isolated peaks are scattered widely over the peak forest plain, and at least 600 m of Devonian limestones lie beneath the surface of the plain. Caves are mainly horizontal galleries and of a water-table type. Peak cluster developed where the vadose zone is thick, usually in the hills at higher altitudes than the plain. Infiltration of rainfall in the peak cluster and dissolution of the rock is much more vertically directed. Caves in the peak cluster are steeply dipping and often narrow. Thus, the two types of karst landscape gave rise to different types of hydrology. On the peak forest plain the karst aquifer is of an open and interconnecting type with some amount of homogeneity but still basically heterogeneous. The seasonal change in aquifer level is less than ten times from wet to dry season. The well completion rate on the plain is almost 60%. In the plain, east of Guilin, there is quite a rapid circulation of water; many boreholes have been constructed with some adverse effects (with many collapses) upon the cones of depression and the water table (Fig. 82 and Photo 36.). In the peak cluster (fengcong) area, the aquifers are confined and tight and more difficult to develop; the well completion rate in the peak cluster is 20%, i.e. the aquifer is heterogeneous. Discharge varies up to 100 times from dry to wet season. Thus, in order to understand the behaviour of the karst underground water, it is essential to

Photo 36. Collapse caused by overpumping water, Guilin area

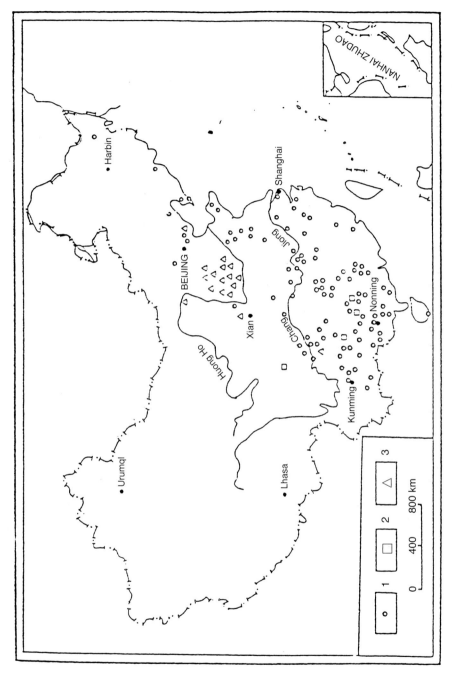

Fig. 82. Distribution of karst collapses in China. *1* Man-induced collapse; *2* natural collapse; *3* palaeocollapse breccia pipe

understand the development of the karst, as the two are completely related and interconnected (Yuan Daoxian 1988).

11.5 The Chemical Quality of the Karst Water

The total mineralization of the karst water in China is below 500 mg/l with a pH of 6.5–8. The quality of the karst water in all parts of China is generally good for domestic purposes. Table 29 gives some of the hydro-chemical features of the different karst areas. The hydrochemistry of the waters reflects the characteristics of the rainfall and surface water inputs and the chemical reactions which occur within the system. In general, carbonate hardness increases from the cooler regions to the warmer ones, as a result of increased vegetation and soil activity and increased pCO_2. Geological factors probably play the major part in determining the chemical characteristics. The water in North China has a much higher SO_4 content which is due to the abundance of gypsum in the Ordovician limestones (Liu Zaihua). The presence of magnesium (MgO) in waters in both North and South China (as at Luizhou) is probably related to the widespread outcrops of dolomites in those areas. Allogenic waters that recharge the karst systems will also change the hydrochemical character of the underground water.

The decline in pCO_2 with altitude is illustrated by a traverse from Sichuan to Tibet, measured in 1987, from 1000 m a.s.l. to over 5000 m at the Tanggula pass (Fig. 66, from Zhang Dian 1991). The effects on conductivity of the waters are also discussed by Zhang Dian. Measurements of soil air in limestone fissures and cave air at different altitudes in Tibet are also shown and indicate relatively high pCO_2 present in the short caves in Tibet (Fig. 67; Sweeting et al. 1991). In other parts of China, soil CO_2 values may be up to 0.08%.

The temperature of the karst water in general again increases with decrease in latitude. It is, of course, modified by relief – for instance some karst springs on the Shanxi plateau have a lower temperature than those at the same latitude on the North China plain. The temperature of spring waters in Tibet are mostly around 0°C – except for those that are from deep tectonic sources which are warm or hot springs.

As we have already seen, large tracts of karst limestones occur in the western arid parts of China – in Xinjiang, Qinghai and in Tibet. The karst hydrogeology in these regions depends on the desert and permafrost hydrological principles. In the dry northwestern desert areas where mean annual rainfall is less than 100 mm, accumulation of water occurs only by infiltration from ephemeral streams (from snow and rainwater falling on the mountains) into the desert fans and basins. Desert crusts of evaporite minerals occur around the lakes. In Qinghai and in Tibet, rainfall may be 300–400 mm but the cold conditions mean that permanent or seasonal permafrost occurs in these areas. Runoff and water circulation are thus limited to the surface rivers and

Table 29. Hydrochemical features of major karst regions in China. (Yuan Daoxian et al. 1991)

Locations	Lat.	Water temp. (°C)	pH	Ca (mg/l)	Mg (mg/l)	Na + K (mg/l)	HCO_3^- (mg/l)	SO_4^- (mg/l)	Cl (mg/l)	TDS
Yichun, Heilongjiang	48°N	8	8.67	35.33	13.23	5.98	172.07	7.00	4.95	239.56
Beijing	40°N	15	7.9	67.53	26.09	13.03	246.51	69.64	11.7	434.5
Baode, Shanxi	39°N	13	7.6	50.3	20.76	27.16	253.54	28.81	14.18	394.75
Jinan, Shandong	36.5°N	18	7.7	70.14	12.15	15.3	225.77	19.21	20.95	363.52
Guilin, Guangxi	25°N	19	7.44	70.94	2.92	1.84	209.91	8.65	4.25	298.51
Liuzhou, Guangxi	24°N	21	7.4	62.12	15.56	4.3	251.39	6.00	6.74	345.11

Conclusion   241

Table 30. Comparisons of karst springs in China with those in former Jugoslavia. (Report of the Institute of Hydrogeology and Engineering Geology 1978–1980. Ministry of Geology and Mineral Resources)

	S China Liulongdong Spring	S China Disu Spring	N China Niangziguan Springs	Ombla Spring	Ljubljanica Springs
Maximum discharge (m³/s)	>74.8	390	16.0	150–165	131.7
Minimum discharge (m³/s)	10.5	4	10.0	4	4.25
Average discharge (m³/s)	23.8	33	13.7	15.2	38.60
Catchment area (km²)	approx. 900	1050	1840	1500	1108

to the active layer in the warm season. Any karst water is normally locked up as ice beneath the surface. In Tibet because of the present active tectonics, there are many hot springs – particularly in the N-S grabens. When these springs occasionally pass through karstic strata they dissolve great quantities of $CaCO_3$ – probably because the water contains not only a great deal of CO_2 but also SO_4 and H_2S as well. As this water reaches the surface, large quantities of $CaCO_3$ tufa is deposited, as at the springs near Amdo. Most Chinese efforts regarding karst hydrogeology are concentrated in the populous eastern parts where careful studies of the exploitation and preservation and storage of the karst water supplies are of greatest importance. In the light of the continuous expansion of Chinese cities (as industrialization and tourism, for example, develop), it is even more necessary to understand karst hydrogeology to prevent both over-pumping and pollution. Lu Yaoru and Liu Fucan compared large karst springs in former Jugoslavia with those in China (see Table 30).

11.6 Conclusion

Karst hydrogeology and karst water are important aspects of karst study in China and there are many detailed records on the behaviour of springs and underground water. Chinese scientists have used this knowledge for practical use, but no great attention has been given to theoretical concepts as in the West. The arguments about the existence or non-existence of water tables, for instance, which figure in Western European literature in karst hydrogeology are not part of Chinese work in this field (Sweeting 1993). This attitude enabled Chinese hydrogeologists to engage in practical karst water schemes long before similar projects were tackled in the West.

12 The Position of China in World Karst Studies

As we have shown, karst studies in China developed separately from the West. Many of concepts are similar in both areas, but because of differences in geology, physiography, climate and in culture and land exploitation, Chinese karst is quite distinctive. It is difficult to compare karstification in different parts of the world since the geological, physiographic and climatic influences are never the same, but certain observations can be made.

12.1 Geological Influences

Karst in China, though widely distributed throughout the stratigraphical column, occurs mostly in the Palaeozoic, except in Tibet where it is in the Jurassic. Many limestones are even older and were formed in the Proterozoic. Such limestones because of their age underwent deep burial, compaction and diagenesis; they are now often sparitic, relatively impermeable, and non-porous and possess considerable mechanical and tensile strength. Such limestones gave rise to strong, usually rugged and steep landforms. By comparison the karsts in Dinaric and Mediterranean Europe formed in Mesozoic limestones (mainly Jurassic and Cretaceous, though some karsts are in Triassic limestones). The Dinaric and Mediterranean limestones are micritic and more porous and subject to rather less diagenesis; they can be, however, mechanically strong. In the Caribbean areas and in Borneo, where tower and cone karst is well developed, the limestones are of Tertiary age (mainly Eocene and Miocene). The Tertiary Caribbean limestones can be porous (up to 20–40%), and are weak compared with those in China; though in Sarawak (Borneo) Tertiary limestones are strong. In peninsular Malaysia, the well-known tower karst in the Ipoh area developed in Carbo-Permian limestones – similar to those in China. The steep towers and cones of the karst where they developed in Guanxi and in Guizhou formed in hard, sparitic, thick-bedded Devonian and Carboniferous limestones.

In western Guizhou in the Shuicheng area, slopes of the cone karst are 46°–48° and often more, but in the Cockpit country of western Jamaica in the Caribbean, slopes of cones in the White Miocene limestones are about 30°–35° (Sweeting 1958). In the Dinaric karst, cone slopes are also 30°–35°; in the Dinaric karst collapses along fracture and joint planes are also important (as at

Skocjan), but generally the Mesozoic limestones do not have the strength of the Devonian sparites of Guilin. A further contrast is provided by the relatively soft limestones in S Mexico (Yucatan) and Florida of Neogene age which have not suffered compaction or much diagenesis; they gave rise to subdued but not rugged relief.

The Ordovician and Proterozoic limestones of northern China, although mechanically strong, are thinner bedded than the massively bedded upper Palaezoics of the south. In N China both Ordovician and Proterozoic limestones formed steep cliffs when subjected to uplift and dissection, but the thin beds and the many joint planes made them susceptible to frost weathering which is important in N China.

The great thicknesses of karst limestones in China are interrupted by sequences of clastic rocks, such as sandstones and shales. These clastic rocks affect the distribution and form of the karst to a high degree. This is seen in the Guilin karst where the Devonian sandstones control the development of the landforms, particularly the fenglin basins (poljes) and the original fluvial valleys. Similarly, dolomitic beds are highly significant in both landform development and karst hydrology. In these respects, the karsts particularly in southern China can be compared with, and are similar to, those in former Yugoslavia. In Croatia, beds of clays and silts, known as *flysch*, interrupt the limestone beds and together with the marked NW-SE overthrusting control the formation of the poljes. Dolomitic beds, in the Dinaric karst, as in China, gave rise usually to gentler slopes and a different karst hydrology (Herak and Stringfield 1972).

The tectonic history of the limestone affects the nature of the jointing and fractures – the older limestones have normally experienced a greater number of tectonic episodes. Strong jointing and massive bedding and fractures are important in the movement and flow of underground water, as is seen in the Rongxian limestones of Guilin and the Urgonian limestones in Dinaric karst. The proximity of the S and W Chinese karst areas to the recent and rapid uplifts of the Himalayas and Tibet means that these areas were affected by neotectonics almost more than any other place in the world. Only in the high mountain karsts of New Guinea have uplifts been so recent and so rapid; but the New Guinea karsts lack the length of time in their development (Barbary et al. 1991). The karst reliefs of Sichuan, Guangxi, Guizhou and Yunnan have a freshness that is rare in Europe. The Dinaric karsts and karsts in southern France (Alpes Maritimes and the limestone Alps of Vercors and Savoie) have all been affected by their proximity to the Alpine (Miocene) uplifts. However, in the Dinaric karst, the period of the recent major tectonics is older than in Guizhou; evidence of peneplanation is more evident, while near to the Adriatic coasts, evidence of marine planation is found. However, both the Dinaric and the southern and eastern French areas are controlled by tectonic lines, as in S China; this is particularly true of the poljes in all areas. In the Dinaric karst the main outlets of underground water are to the W and SW to the Adriatic basin and follow the hydraulic gradients. Because of the rather longer period of evolution of the karst, the underground connections from swallow holes (ponor) to spring outlets seem to be longer and better established in the Dinaric area compared with those in Guizhou. Where, in North China, the establishment of the hydrographic links has had more time, a large underground network of large basins and springs developed. This may be due to the carbonate-sulphate nature of the Magiagou Ordovician limestones.

In gently upwarped platform regions, long caves and large underground catchment basins can evolve. However, in S China, the tight folds and frequent faulting restrict cave development and catchment basins to tectonic units, and this despite favourable climatic conditions and large rivers. Caves may reach 10–20 km in length (as at Duan and Disu, Figs. 46 and 47) but are not likely to reach the lengths of caves like the 120-km-long Mammoth Cave (Kentucky, USA), formed in Carboniferous (Mississippian) limestones in a gently upwarped stable platform area (Palmer 1981). Layered caves associated with platform upwarping, like the Yaolin cave near Hangzhou, are fairly common in many parts of China. Yaolin cave in Zhejiang is situated on the Yangtse platform (Lin Jinshu et al 1987).

12.2 Physiographical Influences

The influence of tectonics and neotectonics is accentuated by the existence of the great rivers of China. The Chang jiang and Zhu jiang in particular and their tributaries bring down great volumes of water and have been able to cut deep canyons into the karstic terrain. Much of the area cut through by the Chang jiang in the San Xia (Three Gorges) is of limestones; in Guizhou and in Yunnan and western Guangxi, rivers like the Sancha He, Wujiang and Hongshui have cut deep gorges over 1000 m deep into the karstic plateau as a result of the recent uplifts; this produces a deep dissection and rejuvenation of the karst, particularly of the underground hydrographic network and changes the nature of the landforms. Because the rejuvenation proceeds headward up the river gorges, there is a zonation of karst landforms from the canyon edges to the basin divide areas; the older karst landforms are in the basin divide areas, and the younger more rejuvenated forms are in the neighbourhood of the canyons. Such zonation was first observed by Chinese geomorphologists, Yang Huairen and Chen Shupeng. H. Baulig also called attention to a rather similar development in the Grands Causses of the Massif Central, in France, in 1934; the zonation, however, is infinitely clearer in South China, because of the greater uplift and greater vigour of the dissecting rivers.

The effects of the great monsoon rainfall and the strongly seasonal climate on the great rivers can be seen, if the gorges are compared with those in former Yugoslavia and southern France, areas in Europe of tectonic uplift but generally much shorter rivers and usually less rainfall. These gorges in Europe are equally steep and sometimes equally deeply cut, but the effects of fluvial action are much less evident and the catchment areas and the volume of water in the rivers by comparison with the Chinese rivers are relatively small. Examples of such limestone gorges are the Tarn and the Var in France, and the Neretva and Krka in the Dinaric karst and the Tara canyon in Montenegro. In Europe, these areas are also much nearer the sea. There is much less interdigitation of fluvial and karstic erosion in the southern European karst areas. This is largely climatic as well as tectonic. To the writer, one of the most striking features of the karst of South China is the combination of fluvial and karstic denudation, particularly well seen in Longgong and Gebihe karst areas of Guizhou. This is the result of the tectonic uplifts, the highly dissected nature of the relief and the all important part played by the rivers and gullies in karst development. The absence of great rivers despite sometimes quite heavy precipitation (over 3000 mm) in the karsts of Europe may be one reason for the emphasis by Cvijic and his successors (particularly Roglic) on the concept of holokarst (Roglic 1964). This concept had a great effect on karst thinking in Europe, which was to some extent detrimental. The idea of holokarst was that of a karst in which fluvial influences were at a minimum, if not absent, and that an area could not be a "true karst" if it contained such fluvial influences. This concept held up European thinking on karst, as it tended to isolate karst thinking from general progress in geomorphology. This was because of the lack of consideration that in all karsts fluvial and karstic processes interdigitate. It is better, in general, to regard the karst as a variant in the fluvial system rather than an isolated system. This view is probably nowhere better seen than in the karsts of Guizhou; it is not so readily observable in the south European karsts.

12.3 Climatic Influences

Tectonic influences are well seen in China, as are also those of climate, since China possesses a wide range of climatic types. The most important climatic variable is precipitation, in both its amount and intensity. The amount and nature of the rainfall is fundamental, since they influence the volume and rate of runoff from the karst surfaces. Clearly, rock type and steepness of slope also affect runoff, but the results upon a limestone surface of light, drizzly rain in an oceanic temperate climate will be very different from heavy tropical seasonal monsoon downpours, such as those experienced in the southern parts of China. The variations caused by temperature differences are usually less significant, except insofar as they affect the distribution of frost and the amount of vegetation growing upon the limestones and therefore control variations in biogenic carbon dioxide. Corrosion rates based on temperature and CO_2 pressures vary with latitude, altitude, and temperature and are often regarded as important variables in the production of the actual landforms. However, corrosion rates are much less important than the nature of the rainfall and its runoff upon the development of the karst landforms. The intensity of the rain storms which produce a high percentage and a very rapid runoff is one of the main factors responsible for the development of the steep slopes and limestone surfaces in the tropical and subtropical parts of South China. The steep cone-like stone peaks of the fengcong are to this extent climatically controlled.

Comparisons with Bosnia and Herzegovina again are valuable in this respect. In Montenegro, also Mesozoic limestones are at 1500–2000 m and occur in one of the wettest parts of Europe, with an average rainfall varying from 1000–5000 mm year, falling mostly in winter. The latitude of Montenegro is approx. 43°N and the climate is of the Mediterranean type. Rainfall intensities are greater than in western Europe, but not as great as in a tropical monsoonal area. There is in Montenegro a great deal of doline development and not as much surface runoff as in China, even though the slopes are steep. Low cone-like residuals occur between the closely set depressions, but they are not as well developed as the fengcong. Relief like the fengcong probably no longer forms in Montenegro today, though there may be evidence of its formation in the past when the climate in Bosnia and Herzegovina was possibly more tropical. The differences might, however, be due to a weaker lithology in Montenegro and a thicker soil or terra rossa cover, or to a much more recent uplift in this part of the Adriatic.

In North China the amount of rainfall is lower, so that the cracks and fissures in the rocks are able to absorb much of the water, and runoff is distinctly less. Instead of steep gullies formed by rapid runoff, water seeps into fissures and small dolines and closed depressions are formed, giving rise to a more characteristic "temperate" karst. The balance between the runoff and the absorption capacity of the limestones will determine the extent to which drainage lines and closed depressions will form (Smart 1988). Thus, the karsts of North China, in Shandong, and in Liaoning seem today to be more like those of central and western Europe, for instance the Domica area of Slovakia and in Hungary or the Jura mountains in France, or in parts of W Virginia, in the USA. As in central Europe, many of the karsts of N China are not in equilibrium with their present-day climates.

The influences of climate are well seen in the desert karsts of West China, which developed in the folded ranges, like the Kunlun Shan and Tinshan and in the intervening basins like Qaidam. There are no real equivalents in China of desert platform karsts like the Nullabor in Australia or those on the Arabian shield in Arabia. Climate and uplift combine to produce uniquely the cold desert and periglacial karsts of Tibet. In the Minshan and other ranges of W Sichuan, high mountain frost pinnacle karst resembles other relatively dry and high mountain ranges. The calcareous massifs of the western and central Swiss Alps in Europe are much wetter (and snowier) and are more glacierized. The high ranges of W Yunnan and E Tibet have more rainfall and small glaciers, but because of their lower latitude are more heavily vegetated than the European high mountain karsts, and the vegetation and snow lines are much higher. The karsts of the Tibet side of the N Himalayas resemble the karsts in the Rocky Mountains of Canada.

12.4 Influences of the Quaternary Period

Because of the long geological history of the Chinese limestones, they have undergone many vicissitudes – burial, folding and faulting, and re-exposure. Hence, palaeokarst is very common, but is also frequent in central and southern Europe. The climatic changes of the Quaternary period have had important effects upon the karsts of China. The main phases of the Quaternary in N China are known from the loess-palaeosol records discussed in Chapter 1, but the record for S China is still largely unknown.

The main glaciers in China today occur in the Himalayas, in the higher ridges in Tibet and in the high ranges of central Asia, like the Kunlun Shan and the Tienshan. Extensions of these glaciers during the Quaternary were relatively small and in the writer's view there is little evidence for Kühle's belief that the plateau of Tibet was covered by ice to a depth of some hundreds of metres during the Quaternary (Kühle 1987). The Quaternary in much of central Asia was probably fairly dry despite oscillations of humid and dry phases. It is believed also that in north and northeast China the Quaternary was essentially dry. This means that there is very little glaciated karst in China. What there is, is mainly at high altitudes as in the high mountain ranges of the Himalaya like Mt. Zebri, near Dingri. There seems to be hardly any, if at all, glaciated karst at fairly low altitudes, which might be expected from the latitudes of the karst areas of Liaoning and Jilin. There is thus no equivalent in China of the great spreads of lowland glaciated karst such as occur in northwestern Britain and Ireland, and in parts of North America, such as in Anitcosti Island and in the Lake Huron area of Ontario. There are few examples of glaciated limestone pavements in China and those only at high altitudes. This is because the glaciations in oceanic western Europe were much wetter phenomena, accompanied by greatly increased rain and snowfall, unlike the dry conditions that seem to have affected North China during the Quaternary. The effects of the rise of the Tibet plateau intensified the monsoonal conditions in South China, but led to a greater aridity in both N China and in the inland regions such as Shanxi and further west. Thus, lowland glaciated karst is one of the few types of karst not really represented in China. We have seen also how the great loess deposits cover much of the karst in Shanxi, making a subhumid area even more arid. Though there was loess deposition in Europe and in North America, it was on a much smaller scale than in China. The impact of the Quaternary upon South China is still relatively unknown, except that the debris flows and mud flows are now known to have been important (Derbyshire 1983).

12.5 Influences of Land Exploitation

The use of the karst has been quite different in both Europe and China. Karst areas have always been valued for their springs and underground water – both in early history in China (see Chap. 2) and by the Minoans and Greeks in Europe, but other attitudes to karst have been different. The main exploitation problems in the karst areas have to deal with the fluctuating karst water levels and discharge. The Chinese people developed irrigation skills very early and their gifts at managing water extended to their use of karst water. Because of their emphasis on rice cultivation, all the karst lowlands and poljes were managed with that in mind. Attempts have been made to control the periodic flooding in the poljes; ponors and swallow holes have been cemented up; and irrigation channels controlled – all this despite the problems of varying water levels and caves. In addition, underground barriers and power stations have been built in caves and large quantities of water stored underground. In southern Europe, utilization of groundwater in the poljes and plains is much more recent; until the 18th century in Croatia and Bosnia the polje floors were not used for cultivation (Herak and Stringfield 1972). They were left to flood naturally in winter in the wet season as the karst water levels rose and then used as summer pastures when water levels fell. They are more controlled now, but not to the extent that they are in southern and central China. The polje floors, therefore, in China tend to be flatter, emphasizing the demarcation between hillslope and plain.

The attitude of the Chinese to mountains also generally differs from that of the European. The cone and tower slopes have been left unused for agriculture and only cleared and deforested. The Chinese idea of mountains is exemplified by the old fable quoted (in a political sense) by Mao Zedong, but equally applicable to the cultural attitudes to the mountains:

"Next day the Foolish Old Man took his whole family to move the mountains. They went out early and came back late, and with no fear of difficulties they dug at the mountains every day.

An old man known as the Wise Old Man saw them moving the mountains and thought it ridiculous. He asked the Foolish Old Man, 'How can you remove two huge mountains at your age?'

The Foolish Old Man replied, 'Although I will die, there will still be my sons. After they die, there will be my grandsons. We will have more and more people, while the mountains will have less and less stones. As long as we have the determination, we can surely remove the mountains'.

When the Wise Old Man heard this, he had nothing to say."
(Quoted by Stoddart 1978 from Mao Zedong 1965).

As a result, the limestone hills and slopes were deforested, very soil-eroded and became bare stone peaks. Many hills have also been literally removed by quarrying in Guilin and Guiyang. The slopes were never used for animal husbandry or arboriculture. This is a great contrast to the use of limestone mountains in the Alps and the Mediterranean areas with their complex development of Alpine pastures and transhumance which have had a long history since Neolithic times. Since 1900 also there has been much re-afforestation of the limestone hills in Europe as in the classical karst of Slovenia and Trieste. The Chinese have recently become greatly concerned about the lack of use and degradation of the stone peaks in the karst areas and an afforestation programme is beginning. Unfortunately, many of the trees planted are imported varieties of conifers and eucalyptus, which are not very suitable. In parts of Guizhou also, the steep slopes have been used for maize cultivation, which has generally increased soil erosion.

In N China (Shanxi and Hebei), limestones have been quarried for decorative stone. Generally, however, the Chinese never used limestones on any scale for building stone. The classical Chinese buildings (pavilions, temples, mansions etc.) were built of wood with tile or slatted wooden roofs; it is only in the non-Han Chinese areas of the minority peoples that housing makes use of the local stone. This contrasts strongly with the use of limestones by the Egyptians and by the Greeks and Romans in the building of temples, aqueducts and even housing; and of course in the later Mediaeval and modern periods in Europe, for churches and other architectual developments.

These differences in land use between the Chinese and European karsts, again emphasizes the contrasts between the flat lands of the poljes and the steep slopes of the cones and towers – Chinese land exploitation tends to accentuate the natural landscape facets.

The greatest contribution made by the Chinese people to karst has probably been their knowledge of karst hydrology and their use of karst water. Their development of karst springs is unrivalled and they attempted to control the fluctuating water tables in karst poljes long before any such efforts were made in SE Europe. The depth of knowledge of Chinese engineers in the karst has enabled over 5000 reservoirs (mostly medium and small) to be constructed in Hunan, Guizhou and Guangxi alone. There are many studies on the problems of water leakage (Yuan Daoxian and Cai Guihong 1987). Such knowledge has also been used to construct hydroelectric power stations in the karst areas, notably in the deep valleys of Guizhou and in developments along the San Xia (Three Gorges) in the middle reaches of the Changjiang. The dam of the Wujiangdu hydroelectric station in Guizhou is 165 m high and has 214×10^7 m^3 storage capacity (Yuan Daoxian et al. 1991). The measurement of tritium was one of the methods used to estimate leakage from the deep karst zone at Wujiangdu.

China was also one of the first countries to extract oil and gas from carbonate reservoirs. Natural gas has been extracted from carbonate rocks in Sichuan since the Han Dynasty (206 B.C.), but since the early 1950s many oil and gas fields in carbonates have been explored and exploited. The discovery of the "buried palaeopeak" type of oil field at Renqui, Hebei, N China in the 1970s showed a new perspective for exploring oil reservoirs in the karst areas of China (Yuan Daoxian et al. 1991).

These facts illustrate the Chinese preoccupation with the practical and applied aspects of karst science. Emphasis on conceptual and theoretical ideas has been much less than in Western science; however, the exploitation of the karst areas could not have taken place without some theoretical background. Examination of the Chinese ideas has enabled us in Europe and America to re-examine our own concepts and see karst geomorphology in a new and different context. The opening up of the Chinese karstlands to foreign scientists has invigorated the study of karst geomorphology. It has made us question complacent attitudes and given us new lines of work for future collaboration between Chinese and Western karst geomorphologists.

References

Academia Sinica (1981) Geological and ecological studies of the Qinghai-Xizang plateau, 2 vols. Science Press, Beijing

Back W, Hanshaw BB, Van Driel JN (1984) Role of groundwater in shaping the eastern coastline of the Yucatan peninsula, Mexico. In: La Fleur RG (ed) Groundwater as a Geomorphic Agent. Allen and Unwin, Winchester, MA, pp 281–293

Bakalowicz M, Mangin A (1980) L'aquifère karstique. Sa definition, ses characteristiques et son indentification. Mem L Ser Soc Géol Fr 11:71–79

Balázs D (1961) Physische Geographie der Südchinesischen Karstgegend. Academia Scientianuar Hungarica. Institutum Geographicum, Budapest, Publ no 83 (in Hungarian)

Balázs D (1962) Beiträge zur Speläologie des Südchinesischen Karstgebieten. Karszt-es-barlang-kutatás 2, Hungarian Caves and Karst Journal, Budapest, pp 3–82

Balázs D (1973) Relief types of tropical karst areas. IGU Symp on Karst morphogenesis. Szeged Attila Jozsef University, pp 16–32 Szeged, Hungary

Balázs D (1989) Karst és Barlang, 1-II. Hungarian Caves and Karst Journal, Budapest, pp 41–50

Bao Haosheng, LI Chunhua, PAN Ruihong, LI Zhibin (1991) Cave resources and the impact of human activities on cave environment in Yangxian scenic spot of Yixing county, Jiangsu province. J Nanjing Univ (Nat Sci Ed) 27 2:360–369 (in Chinese, English Summary)

Barbary JP, Maire R, Zhang Shouyue (1991) Gebihe '89. Karstologia Mém 4:232 pp

Barbour GH (1930) Taiku deposits and the problem of Pleistocene climates. Bull Geol Soc China 10:71–104

Baulig H (1934) Le plateau central de la France. Librairie Armand Colin Paris

Black D, Teilhard de Chardin P, Young CC, Pei WC (1933) Fossil man in China. Mem Geol Surv China A 11:1–158

Bosák P, Ford DC, Glazek J, Horacek I (eds) (1989) Palaeokarst: a systematic and regional review. Elsevier, Amsterdam 725 pp

Bouillard G (1924) Les grottes de Yunshui cave du Changfang Mtn. Bull Geol Soc China 3, 2:147–152

Bremer H (1981) Zur Morphogenese in den Feuchten Tropen. Verwitterung und Reliefbildung am Beispiel von Sri Lanka. Relief. Boden Palaeoklima 1. X. Borntraeger. Berlin, 196 pp

Bretz JH (1956) Caves of Missouri. Mo Geol Surv Water Resour 2:39

Brook D (1993) The Anglo-Chinese expedition to Sichuan ('92). Caves and caving. Br Cav Res Assoc 59:9–11

Bull PA, Yuan Daoxian, HU Mengyu (1989) Cave sediments from Chuan Shan tower karst, Guilin, China. Cave Sci 16, 2:51–56

Bull PA, Magee AW, Sweeting MM (1990) Tibetan cave sediments; an SEM study of clastic deposits from Tibetan palaeokarst. Carsologica Sinica 9:76–87 (in Chinese and in English)

Burger D (1990) The travertine complex of Antalya/south west Turkey. Z Geomorphol Suppl 77:25–46

Cao Jiadong, Shao Shixiong (1991) Neotectonic movements in China. In: Zhang Zonghu, (ed) The Quaternary of China. XIII Inqua Congr. China Ocean Press Beijing

Chan Man-Suen et al. (1987) An outline survey of the algae and chemical nutrients of the Dianchi lake. Oxford University Freshwater Biology Project, Yunnan, Oxford, 83 pp

Chen Kezao, Bowler JM (1988) A preliminary study on the sedimental features and the palaeo-climatic evolution in Qarhan Salt lake in Qaidam basin. Sci Sin 5

Chen Shupeng (1957) Karst geomorphological map of Seven Star cave in Guilin. Geographic information, vol 1. Scientific Publishing House, Beijing, pp 56–75 (in Chinese)

Chen Shupeng (1965) On the movement of karst water. First National Symposium of Hydrology and Engineering Geology. (in Chinese)

Chen Shupeng (1986) Scientific and epochal contribution of 'Xu Xiake's travels' – verifications in Nujiang and Tengchong area, SW Yunnan. Geogr Res 5, 4:1–11 (In Chinese with English Summary)

Chen Wenjun (1988) The study of Disu underground river system, Du'an county, Guangxi. Proc of the IAH 21st Congr. Guilin, pp 543–551

Chen Zhenpeng (1988) The study of water supply in Jinan under the preservation of springs. Carsologica Sin 4, 1:22–30

Chen Zhiping, Song Lin hua Sweeting MM (1986) The pinnacle karst of the stone forest, Lunan, Yunnan, China; an example of subjacent karst. From new directions in karst. In: Paterson K, Sweeting MM (eds) Proc of the Anglo-French Karst Symp, 1983. Geo Books, Nowicl, pp 597–607

Chui Zhijiu (1979) Palaeokarst on the Qinghai-Xizang plateau. Nature 2.4. Beijing (in Chinese)

Chui Zhijiu (1983) Tibetan geomorphology. Science Press, Beijing, pp 155–166 (in Chinese)

Chui Z, Zheng B (1976) Karst in the Everest area. In the report of science expedition in the Everest area. Science Press, Beijing (in Chinese)

Cotton CA (1948) Landscape as developed by the processes of normal erosion, 2nd edn. Cambridge University Press, Cambridge, 474 pp

Cui Guangzhong (1985) Rapid flow of karst groundwater system in China. Karst water resources. Proc of the Ankara-Antalya Symp, 1985. IAHS Publ 161:619–632

Cuisinier L (1929) Régions calcaires de l'Indochine. Ann de Géogr 38:266–73

Cvijic J (1893) Das Karstphänomen. Penck's Geogr Abh (5)3:217–330

Cvijic J (1924) The evolution of lapies; a study in karst physiography. Geogr Rev 14:26–49

Day MJ (1979) The hydrology of polygonal karst depressions in northern Jamaica. Z Geomorphol Suppl 32:25–34

Derbyshire E (1983) The Lu Shan dilemma. Z Geomorphol NF 27, 4:445–471

Dewey JF, Chen Changfa, Shackleton RM (1988) The tectonic evolution of the Tibetan plateau. Philos Trans R Soc Lond Ser A 327:379–413

Dong Bingwei (1987) Tenglong cave system and its value in development. Carsologica Sin 6, 4:323–327 (in Chinese)

Dreybrodt W (1988) Processes in karst systems. Series in physical environments 4. Springer, Berlin Hedeberg New York, 288 pp

Drogue C, Bidaux P (1992) Structural and hydrogeological origin of tower karst in southern China. Z Geomorphol NF 36, 1:25–36

Eavis A, Waltham AC (1986) China caves. Royal Geographical Society, London

Felber H, Hotzl H, Zöth J et al. (1978) Karstification and geomorphology of AsSulg plateau. In: Al-Sayari SS, Zötl JG (eds) Quaternary period in Saudi Arabia. Springer, Vienna New York pp 166–172

Ford DC (1979) A Review of alpine karst in the southern Rocky mountains of Canada. Nat Speleo Soc Am Bull 41:53–65

Ford DC (1983) The physiography of the Castleguard karst and Columbia icefields area, Alberta, Canada. Arct Alp Res 15, 4:61–71

Ford DC, Williams P (1989) Karst geomorphology and hydrology. Unwin Hyman, London, 601 pp

Ford DC, Fuller PG, Drake JJ (1970) Calcite precipitates at the sole of temperate glaciers. Nature 226 (5244):441–442

Fort M, Dollfus O (1989) Observations et hypotheses geomorphologiques dans les Kunlun et au Tibet occidental. Université Paris VII. Travaux du Laboratoire de Géog Physique 18, 66 pp

Gao Daode, Zhang Schichong, Zhang Jingjual (1986) Karst in South Guizhou, China. Guizhou People's Publishing House, Guiyang, Guizhou

Gao Daode et al. (1988) Karst and karst groundwater in South Guizhou China. Proc 21st IAU Congress, Guilin, pp 532–542

Gellert JF (1962) Der Tropenkarst in Süd China im Rahman der Gebirgsformung des Landes. Verhandlung Deutscher Geographentag 33, East Berlin, pp 376–384

Geng Hong, Cheng Shuyun, Luo Suxiang (1988) Geological excursion guide book to the karst landscape in the Lunan region. Yunnan Prov Bureau of Geology and Mineral Resources, Yunnan province. (in English and Chinese)

Gregory JW (1911) Constructive waterfalls. Scott Geogr Mag 27:537–546

Guan Yuhua, Song Linhua (1984) Salt karst in Qinghai plateau, China. Le Grotte d'Italia (4) XII:337–345

Guan Yuhua, Xu Yaoxian (1981) The salt karst of Qarhan Salt lake, Qinghai. Carsologica Sinica 4. 1 and 2:175–188 (in Chinese, English Summary)

Han Tonglin (1991) In the Quaternary of China. Quaternary geology of the Qinghai-Tibet (Xizang) plateau. Academia Sinica, Beijing, pp 405–440

Handel-Mazetti H (1926) Portraits of nature in SW China. Naturbilder aus Südwest-China. Viennna, 283 pp (in German)

He Caihua (1982) The characters of karst geomorphology in Guizhou. Dept of Geography, Guizhou Normal University, Guiyang

He Caihua (1987) The characteristics of karst geomorphology in Guizhou. International Geomorphology 1986. Part II. Wiley, Chichester, pp 1095–1108

Hedin S (1898) Through Asia, 2 vols. Methuen, London

Herak M, Stringfield VT (1972) Important karst regions of the N hemisphere. Elsevier, Amsterdam 551 pp

Holland CH (1990) The Yangtze platform; a gateway to Chinese geology. Proc Geol Assoc 101, 1:1–17

Hu Bangbo (1991) Xu Xiake, a Chinese traveller of the seventeenth century and his contribution to karst studies. Cave Sci 18, 3:153–157

Huang Wanpo (1977) On the age of the cave faunas of South China. Vertebr Palasiat 17, 4:327–342 (in Chinese)

Jakucs L (1977) Morphogenetics of karst regions. Adam Hilger, Bristol, 284 pp
Jennings JN, Sweeting MM (1963) The limestone ranges of the Fitzroy basin, western Australia: a tropical sand-arid. Bonn Geogr Abh 32:1–60
Kozarski S (1962) Sub-tropical needle karst between Kweilin and Yang-shuo, Kwangsi. Bull Soc Amis Sci Lett Poznan 7:48–70
Kuhle M (1987) Subtropical mountain and highland glaciation as ice age triggers and the waning of the glacial periods in the Pleistocene. GeoJournal 14, 4:393–421
Kuhle M (1991) New results on the Pleistocene glaciation of Tibet and the onset of ice ages, Abstr XIII INQUA Congress, Beijing, 178 pp
Lee JS (1933) (LI Siguang) Quaternary glaciation in the Yangtze valley. Bull Geol Soc China 13:15–62
Lehmann O (1926) The geographical results of the travels (Handel-Mazetti) through Kweischou (S China). Akad Wiss (Vienna) Math Naturwiss Denkschr 100:77–99
Lehmann H (1936a) Morphologische Studien auf Java. Geogr Abh 9 (Ser 3):1–114
Lehmann H (1936b) translated 1981 in Sweeting MM (1981) Karst geomorphology. Benchmark Papers in Geology, vol 59. Hutchinson Ross, Stroudsburg, 327 pp
Li Chuanmo (1985) Analysis on karst resources and preservation of famous springs in Jinan. Carsologica Sin 4, 1:31–39
Li Datong et al. (1985) The map of soluble rock types in China, 1:4 000 000. Cartographic Publishing House Beijing (in Chinese)
Li Yanxian (1981) On the subdivisions and evolution of the Quaternary mammalian faunas of South China. Vertebr Palasiat 19, 1:67–76 (in Chinese)
Li Siguang (Lee JS) (1939) The geology of China. Thomas Murby, London, 528 pp
Lin Jinrong (1992) A discussion on the development history of tropical karst and its palaeoenvironment in Guilin. In: Yuan Daoxian (ed) World Karst Correlation. Carsologica Sin 16:110–117
Lin Jinshu (1982) An analysis of the palaeogeographic elements of karst development in the Wumin basin. Acta Geogr Sin 37, (2)
Lin Jinshu Zhang Yaoguang, Huang Yunlin (1987) A preliminary study of development processes and dynamic conditions in the Yaolin cave in E China. International Geomorphology, Part II. John Wiley, New York, pp 1143–1150
Lin Z, Wu X (1981) Climates of the Xizang-Qinghai plateau before and after uplift. In: studies on the period, amptitude and type of uplift of the Xizang-Qinghai plateau. Science Press, Beijing (in Chinese)
Liu Jiming et al. (1991) Investigation of the vegetation character of three typical karst landforms in Maolan. The Management Section of the Maolan National Nature Reserve. Libo County, Guizhou Province (for the IGCP Project 299)
Liu Tungsheng (1988) Loess in China, 2nd edn. China Ocean Press, Beijing; Springer, Berlin Aeidelberg New York. XX + 224 pp
Liu Zaihua (ed) (1991) Guide book for field excursions. Karst of the inner plate region with monsoon climate. Institute of Karst Geology, Guilin, pp 19–22
Liu Zaihua et al. (1992) Influence of hydrodynamic conditions and water chemistry on the formation of tufa in Huanglong, Sichuan. In: Yuan Daoxian (ed) World Karst Correlation Carsologica Sin 16:79–91
Liu Zechun (1985) Development and filling of caves in Dragon Bone hill at Zhoukoudian, Beijing. International Geomorphology, 1986. Part II. John Wiley, New York, pp 1125–1141
Lu Yaoru (1980) Preliminary comparison of the basic features of karst development

between China and Jugoslavia. Institute of Hydrogeology and Engineering Geology, Beijing, pp 117-121 (in Chinese, English Summary)

Lu Yaoru (1986) Karst in China. Geological Publishing House, Beijing (in Chinese with many photographs)

Lu Yaoru (1987a) Karst geomorphological mechanisms and types in China. International Geomorphology 1986. Part II. John Wiley, New York pp 1077-1093

Lu Yaoru (1987b) Water resources in karst regions and their comprehensive exploitation and harnessing. International Geomorphology 1986, Part II. John Wiley, New York, pp 1151-1167

Lu Yaoru (no date) Karst geomorphological mechanisms and types in China. Beijing, 22 pp

Lu Yaoru, Lin Jinshu, Song Linhua (1972) The development of karst in China and some of its hydrogeological and engineering geological conditions. Acta Geol Sin 22:1-19

Lui T, Ding M (1980) Palaeoclimatic records in the loess of China and their reflection of the ancient climate evolution. Scientific Papers on Geology for International Exchange, Beijing, pp 77-82

Lui Ze chun (1985) Sequence of sediments at locality I in Zhoukoudian and correlation with loess stratigraphy in northern China and with the chronology of deep sea cores. Quat Res 23:139-153

Mao Zedong (1965) Selected works of Mao Zedong. Foreign Languages Press, Beijing, 312 pp

Masschelein J, Zhang Shouyue (eds) (1990) Teng Long Dong; the longest cave of China. Report of the 1st Belgian-Chinese speleological exped 1988. Belgian – Chinese Karst and Caves Assoc, Liège, 46 pp

Meyerhoff AA, Kamen-Kaye M, Chin Chen, Taner I (1991) China; stratigraphy paleogeography and tectonics. Kluwer, Dordrecht

Needham J (1954) Science and civilisation in China, vol 1. Cambridge University Press, Cambridge, p 62

Needham J (1959) Science and civilisation in China, vol 3. Cambridge University Press, Cambridge

Nicod J (1992a) Splendeur et problèmes des karsts de la Chine du Sud. Ann Géogr 567:595-601

Nicod J (1992b) Les karsts sous couverture. Cuad Sec Hist 20:165-185

Nicod J, Fabre G (1982) Modalités et rôle de la corrosion crypto-karstique dans les karsts mediterranéans et tropicaux. Z Geomorphol 26, 2:206-224

Osmaston H (1992) Tropical type tors and tafoni in Tibet and Ladakh. Paper presented at 7th Himalaya-Karakoran-Tibet Worksh, Oxford, April 1992. Abstr 119

Palmer AN (1981) A geological guide to Mammoth Cave National Park. Zephyrus Press, Teaneck, NJ

Penck A (1924) Das Unterirdische Karst. Receuil de Travaux offert à J. Cvijic Belgrade. Spec Vol, University Belgrade

Plummer LN, Wigley TML, Parkhurst DL (1978) The kinetics of Calcite dissolution in CO_2 water systems at 5°C to 60°C and 0.0 to 1.0 atm. CO_2. Am J Sci 278:179-216

Pohl ER (1955) Vertical shafts in limestone caves. Nat Speleol Soc Am Occ Pap 2:

Ren Mei E (1959) About Xuxiake's travels. In selected works of ancient geographical classics in China. Science Press, Peking

Ren Mei E (1984) Xu Xiake's contribution in the study of karst. Acta Geogr Sin 39, 3:252–258
Ren Mei E, Liu Ze Chun (1982) Development of the Peking Man's cave in relation to early man at Zhoukoudian, Beijing. In: Quaternary Geology and Environments of China. China Ocean Press, Springer, Berlin Heidelberg New York, pp 137–142
Ren Mei E, Wang Kezao, Liu Zechun (1981) Evolution of limestone caves in relation to the life of early man at Zhoukoudian, Beijing. Sci Sin XXIV, 6:843–851
Ren Mei E. et al. (1982) The morphological characteristics of karst in China. Nanjing University Paper, pp 1–24
Ren Mei E, Yang Renzhang, Bao Haosheng (1985) An outline of China's physical geography. Foreign Languages Press, Beijing, 471 pp
Roglic, J (1939) Beitrag zur Kenntnis der Karstformen in den Dinarischen Dolomiten. Hrv Geogr Glasnik, University of Zagreb
Roglic J (1964) Karst valleys in the Dinaric karst. IGU Symposium Erdkunde 18(2):113–116
Roglic J (1974) Les caractères spécifiques du karst Dinariques. Mém Doc 15:269–278
Ru Jinwen (1981) Crown rock ground water system. Sci technol Karst 1,2:17–24, 32–35 (in Chinese)
Ru Jinwen, Zhu Dehao, Liu GongYu (1991) Karst in Guilin. A guidebook. Institute of Karst Geology. Press of the University of Science and Technology of China, Beijing, 77 pp
Ruxton BP, Berry L (1957) Weathering of granite and associated erosional features in Hong Kong. Bull Geol Soc Am 68:1263–1292
Salomon JN, Astruc G Jr (1992) Exemple en zone temperée d'un paleocryptokarst tropical exhumé (la Cuvette du Sarladais). In: Salom on JN (ed) Karst et Evolutions climatiques. Presses Universitaires de Bordeaux, Bordeaux, pp 431–447
Shackleton NJ, Opdyke ND (1973) Oxygen isotope and paleomagnetic stratigraphy of equatorial pacific core V28–238. Oxygen isotopes temperatures and ice volumes in 10^5 year and 10^6 year scale. Quat Res 3:39–55
Shackleton RM, Chang Chengfa (1988) Cenozoic uplift and deformation of the Tibetan plateau; the geomorphological evidence, Philos Trans R Soc Lond Ser A 327:365–377
Shi Yafeng (1992) Glaciers and glacial geomorphology in China. Geomorphol Suppl 86:51–63
Shi Yafeng, Liu Turgseng (1992) Last glaciation and maximum glaciation in the Qinghai-Xizang (Tibet) plateau: a controversy to M. Kuhle's ice sheet hypothesis. Z Geomorphol Suppl 84:19–35
Shi Zhenbing (1988) Researches on tufa in the north western Sichuan. Thesis, Nanjing University, Dept of Geography, 66 pp (in Chinese, English Summary)
Silar J (1963) Zur Morphologie und Entwicklung des Kegelkarstes in Süd-China und Nord-Vietnam. Petermans Geogr Mitt 107:14–19
Silar J (1965) Development of tower karst in China and North Vietnam. Natl Speleol Soc Am Bull 27:35–46
Smart CC (1988) A deductive model of karst evolution based on hydrological probability. Earth Surface Processes Landforms 13, 3:271–288
Smart P, Waltham AC, Yang Mingde, Zhang Ying-jun (1986) Karst geomorphology of western Guizhou. Cave Sci (BCRA) 13, 3:89–113

Song Linhua (1980) Some characteristics of karst hydrology in Guizhou plateau, China. Institute of Geography, Academia Sinica, Beijing
Song Linhua (1981) Progress of karst hydrology in China. Prog Phys Geogr 5, 4:563–574
Song Linhua (1986a) Karst geomorphology and subterranean drainage in South Dushan, Guizhou province, China. Cave Sci 13, 2:49–63
Song Linhua (1986b) Origination of stone forests in China. Int J Speleol 15:3–13
Song Linhua, Liu Hong (1989) Cockpit karst and Geological structures in South Yunnan, China. 3rd Multidisciplinary Conf on Sinkholes, Florida, pp 377–384
Starkel L, Gregory KJ, Thornes JB (1991) Temperate palaeohydrology. John Wiley New York 568 pp
Stein A (1912) Ruins of desert Cathay, 2 vols. Macmillan, London
Stoddart DR (1978) Geomorphology in China. Prog Phys Geogr 2, 2:187–236
Sun, Sian Qing (1944) Glacial features in NW Kuangsi. Bull Geol China 24:105–113
Swann PC (1956) Chinese paintings and the Chinese landscape. Geogr Mag 28:603–610
Sweeting MM (1958) The karstlands of Jamaica. Geogr J 124:184–199
Sweeting MM (1978) British geographers in China, Geogr J 144, 2:187–207
Sweeting MM (1990) The Guilin karst. Z Geomorphol Suppl 77:47–65
Sweeting MM (1993) Reflections on the development of karst geomorphology in Europe and a comparison with its development in China. Z Geomorphol Suppl 93:127–136
Sweeting MM, Bao Haosheng, Zhang Dian (1991) The problem of palaeokarst in Tibet. Geogr J 157, 3:316–325
Sweinfurth U, Sweinfurth-Marby H (1975) Exploration in the eastern Himalayas and the river gorge country of S.E. Tibet. Francis (Frank) Kindom-Ward (1885–1958). An annotated Bibliography with a map of his Expeditions, Steiner, Vienna
Tan Ming (1992a) Mathematical modelling of catchment morphology in the karst of Guizhou, China. Z Geomorph 36, 1:37–51
Tan Ming (1992b) On the origin of cone karst and its morphology. In: Yuan Daoxian (ed) World Karst Correlation. Carsologica Sin 16:118–120
Teilhard de Chardin P, Young CC, Pei WC (1935) The Cenozoic sequence formations of Kwangsi and Kwangtung. Bull Geol Soc China 14:179–210
von Richthofen F (1877–1912) China; Ergebnisse eigener Reisen und darauf gegründeter Studien, 5 vols, 2 Atlases. Berlin
von Wissmann H (1954) Der Karst der Humiden Heissen und Sommerheissen Gebiete Ost-Asiens. Erdkunde 8:122–130
Waltham AC (1984) Some features of karst Geomorphology in South China. Cave Sci (BCRA) 11, 4:185–198
Waltham AC (1991) Limestone karsts of the Annapurna region, Nepal Himalayas. Trans BCRA 18, 2:99–104
Wang Fakung, Ma Fengshan (1989) Karst and karstic water in Taiyuan and Huoxian region. Institute of Geology, Academia Sinica, Beijing, 10 pp
Wang Fubao, Fan CY (1987) Climatic changes in the Qinghai-Xizang (Tibetan) region of China during the Holocene. Quat Res 28:50–60
Wang Kejun et al. (1988) Quaternary glaciation in the Guilin area. Dedicated to the Centennial of J.S. Lee. Chongqing Publishing House, Chongqing, 107 pp (in Chinese, with English Summary)
Wang Xunyi (1986) U Dating and $\delta^{18}O$, $\delta^{13}C$ features of speleothems in Mao Mao Tou (Big cave), Guilin. Kexue Tongbao 31, 12:835–838

Weng Jintao (1987a) Karst and carbonate rocks in Guilin. Chongqing Publishing House, Sichuan (in Chinese, English Summary)
Weng Jintao (1987b) Rb-Sr dating of illite in red bed series. Kexue Tongbao 32, 13:5–10
Williams PW (1972) Morphometric analysis of polygonal karst in New Guinea. Bull Geol Soc Am 83:761–796
Williams PW (1978) Karst research in China. Trans Br Cave Res Assoc 5(1):29–46
Williams PW (1983) The role of the subcutaneous zone in karst hydrology. J Hydrol 61:45–67
Williams PW (1987) Geomorphological inheritance and the development of tower karst. Earth Surface Processes Landforms 12:453–465
Williams PW et al. (1986) Interpretation of the palaeo-magnetism of cave sediments from a karst tower at Guilin. Carsologica Sin 5(2):113–126
XI Devin (1991) Introduction to the karst hydrogeological conditions of Jinan spring region and the research on the springs protection and water supply. Shandong Geo-Engineering Prospecting Institute, Jinan, Shandong
Xiong Kangning (1985) Morphometric analysis and evolutional regularity of fenglin landscape in the Shuicheng area. Masters Thesis, Guizhou Normal University, Department of Geography
Xiong Kangning (1992) Morphometry and evolution of fenglin karst in the Shuicheng area, western Guizhou, China. Z Geomorphol 36, 2:227–248
Xu Hongxu (1642) Xu Xiake Youyi (in classical Chinese) diary of the travels of Xu Xiake. Shanghai Guji Chubanshe, Shanghai (reprinted 1982)
Yan Qingtong, (1982) Karst of the stone forest. (unpublished manuscript)
Yang Hankui, Zhu Wen Xiau, Huang Renhai (1984) The feature of karst waterfalls in Yunnan-Guizhou plateau of China. Carsologica Sin 1984:89–96. (in Chinese, English Summary)
Yang Hankui, Huang Renhai, Zhu Wenxian (1986) Tufa deposition facies and types. From the Geology of Guizhou. No 1, pp 47–56 Guizhou Unviersity Guiyang
Yang Mingde (1982a) The geomorphological regularities of karst water occurrences in the Guizhou Plateau. Carsologica Sin 1, 2:(in Chinese)
Yang Mingde (1982b) Texture and evolution of karst geomorphology in Guizhou Plateau. Guizhou Science Progress, Guiyang
Yang Mingde, Zhang Ying jun, Xiong Kangning (1988) A study on the karst environmental in Guixhou. Guizhou Society of Environmental Sciences, Guiyang, 107 pp (in Chinese, English Summary)
Yang Zunyi, Chen Yugi, Wang Hongzhen (1986) The geology of China. Univ Press, Oxford, 338 pp
Yu Jinbiao (1982) The period of stone forest origin and analysis of palaeo-geographical environment in Lunan county, Yunnan. In: Luizhou karst, geomorphology and speleology Symp
Yuan Daoxian (1980) A brief account of the karst geology of Guilin Institute of Karst Geology, Guilin, 16 pp
Yuan Daoxian (1981) A brief introduction to China's research in karst. Institute of Karst, Guilin, 35 pp
Yuan Daoxian (1985a) On the heterogeneity of karst water. Karst water resources. Proc of the Ankara-Antalya Symp 1985. IAHS 161:281–292

References

Yuan Daoxian (1985b) New observations on tower karst. Proc 1st Int Conf on Geomorphology, Manchester, 676 pp

Yuan Daoxian (1988) Paper prepared for hydrogeology of selected karst regions of the world. In: Bock W, Paloc H (eds) Karst and karst water in China. Int Assoc Hydrologists, Beijing

Yuan Daoxian, Cai Guihong (1987) The science of karst environment. Chongqing Publishing House, Chongqing, China, 332 + 24 pp (in Chinese, English Summary)

Yuan Daoxian, Drogue C et al. (1988) Report on the Guilin karst hydrogeology experimental site. Private Report to Guilin, Institute of Karst

Yuan Daoxian, Hu Mengyu, Bull PA, Sweeting MM (1985) Aspects of the Quaternary deposits around Guilin, Guangzi, South China. Carsologica Sin 8:267–279 (in English and in Chinese)

Yuan Daoxian, Zhu Dehao, Zhu Xuewen et al. (1991) Karst of China. Geological Publishing House, Beijing, pp 1–124, p 158

Yuan Daoxian, Weng Jeitao, and Zhu Xuewen, Glossary of karst terms in Chinese. Institute of Karst, Guilin

Zhang Dachang (1988) Hydrological conditions and water resources in the northern part of Mt. Xishan, Taiyuan, China. Academia Sinica, Beijing (unpublished manuscript)

Zhang Dian (1991) Evolution of Tibetan karst and landforms. PhD Thesis, Dept of Geography, University of Manchester

Zhang Jie (1987) The karst characteristics and its formational condition; analysis of the Jiuzhaigou district. Thesis Dept of Geography, Nanjing University, 82 pp (in Chinese, English Summary)

Zhang Shouyue: Academia Sinica (1979) Karst group, Academia Sinica of China. Karst study in China. Science Press, Beijing (in Chinese)

Zhang Shouyue (1984) The development and evolution of Lunan stone forest. Carsologica Sin 3, 3:78–87

Zhang Shouyue (1989) Palaeokarst of China. In: Bosák P, Ford D, Glazek J, Horácek I (eds) Palaeokarst Elsevier, Amsterdam 297–311

Zhang Yingjun, Mo Zhongda (1982) The origin and evolution of Orange fall (Huanggoushu). Acta Geogr Sin 37:23–41 September (in Chinese, English Summary)

Zhang Yingjun, Zhang Dian (1984) The types, distribution and evolution of Guizhou caves. Guizhou Science Progress. The Science and Technology Association of Guizhou, pp 169–183

Zhang Yingjun et al. (1985) Applied karstology and speleology. Guizhou People's Press (in Chinese, English Summaries)

Zhang Zemin (ed) (1988) Guide to excursion route C (Shijcazhuang – Beijing). 21st Congress, IAH, Guilin 1988

Zhang Zhigan (1980a) Karst types in China. GeoJournal 4. 6:541–570

Zhang Zhigan (1980b) Sulphate-carbonate karst in the Majiagou limestone in the Niangziguan region, Shanxi province. Institute of Hydrogeology, Hebei

Zhang Zhonghu (ed) (1991) In the Quaternary of China, 13th Inqua Congr Beijing. China Ocean Press, Beijing, V + 575 pp

Zhao Shusen et al. (1988) The U-series ages of speleothems in karst caves in eastern China. Proc of the IAH 21st Congr, Part I, Guilin, Guangxi, pp 241–247

Zhao Songqiao (1986) Physical geography of China. Science Press, Beijing and John Wiley, New York, 209 pp

Zhou Shiying, Zhu Dehao, Lao Wenke (1988) Calculation of karst denudation rate in peak cluster depression in Guilin area. Carsologica Sin 7, 1:72–80

Zhou Xuansen (1981) The Yaolin cave in Zhejiang and its deposits, J Hangzhou Univ 8, (1):4–10

Zhou Zhengxian (1986) Scientific survey of the Maolan karst forest Guizhou People's Publishing House, Guiyang, Guizhou Province 386 pp (in Chinese, English Summaries)

Zhu Dehao (1982) Evolution of peak cluster depressions in the Guilin area and morphometric measurement. Carsologica Sin 2, 2:127–184 (in Chinese)

Zhu Fengjun (1987) The dissolution kinetics of carbonate rock fissures and its application in karst research of Lancuen spring district. Thesis Institute of Geology, Academia Sinica, Beijing, 96 pp

Zhu Xuewen (1981) Karst and karst water in East Sichuan. Karst Institute, Guilin, pp 84–115 (in Chinese)

Zhu Xuewen (1988a) Guilin karst. Shanghai Scientific and Technical Publishers, Guilin, 188 pp

Zhu Xuewen (1988b) Guide to excursion route no 2. Kunming, 21st IAH Congress. Chinese Committee for IAH Geological Society China, Guilin

Zhu Xuewen (1990) Fenglin Karst (tower and cone karst) in China. Institute of Karst Geology, Guilin pp 1–16

Zhu Xuewen, Zhu Dehao, (1988) Karst caves in Guilin. Geological Publishing House, Beijing, 249 pp (in Chinese, English Summary)

Subject Index

Aeolian
　landforms 200
　sediments 215
Ailuropoda-stegadon 73
alpine karst
　see high altitude karst
allogenic systems 65, 113, 139, 140, 148, 165, 166, 177, 236
altiplanation terrace 207
anhydrite 162
antecedent river 197
aquifers 20, 76, 95, 124, 146, 166, 179, 205, 237
　heterogenous 38, 234–235
　homogenous 76, 235
autogenic system 113

Baotu springs (Shandong) 167
bare karst 15–16, 94
barytes 41
bauxite 41, 162
biokarst 183, 194, 218
blind valley 121
Bohai 56
boreholes 179, 230, 234, 237
breakdown 215
breccia 62, 88, 162
Brunhes Normal 171
building stone, limestone 249
buried karst 1, 15–16, 55, 56–57, 151, 155, 159, 224

calcite 40, 186, 190, 192
Calcium carbonate deposition 75, 155, 158, 187, 190
canyon
　see gorge
carbon-14 dating 64, 75, 117, 155, 160, 177, 190

carbon dioxide 21, 74, 75, 83, 116, 131, 148, 183, 194, 201, 202, 239
carbonate, rocks 1, 11–19, 24–25, 66, 152, 197
calcrete 55
caves 23, 35, 37–38, 54, 65, 68–70, 95, 113–115, 119, 150, 159, 170–174, 177, 214, 220
　Chuangshan cave (Guilin) 78
　Da yan cave (Guilin) 75
　Dingri caves (Tibet) 218
　Elephant Trunk Hill cave (Guilin) 72
　Guanyan cave (Guilin) 85–86
　Lingshan cave 149
　Lotus Throne cave (Xingping) 74
　Longgong cave (Guizhou) 109–112
　Maomaotou cave (Guilin) 75
　Moon cave (Yangshuo) 77
　Mt. West caves (Tibet) 213
　Mt. Zebri caves (Tibet) 213
　Nan cave (Guilin) 87
　Peking Man's cave (Xishan) see Peking Man
　Reed Fute cave (Guilin) 72
　Seven Star cave (Qixing Yan) 33
　Shanhu cave (Shandong) 165
　Shanjuan cave (Changjiang) 148
　Tenglong cave (Hubei) 146
　Tiandong cave (Qinling) 155
　Tunnel cave (Guilin) 73
　Wanhuanyan cave (Hunan) 148
　Yaolin cave 149, 244
　Yin Hill cave (Guilin) 70
　Zhijin cave (Guizhou) 113
cave coral 215, 217
chamber, cave 72, 113, 170, 219
Changjiang (Yangtze river) 8, 11–12, 21, 24, 33, 49, 52, 139, 144–150, 183, 244

chert 19, 128
chlorite 217
chuans 157
clastic cave sediments 75, 76, 78, 87, 148, 172–174, 214–215
climate
  effects of 21–23, 33, 46, 120, 153, 154, 196, 245–247
  change 23–30, 118, 218, 247
closed depression 42–43, 45, 53, 100, 109, 114, 121–123, 125, 131, 144, 146, 159, 228, *see also* doline
collapse 14, 162, 171, 225, 237
  see also doline and closed depression
conduit flow 116, 142, 204, 223, 225
cone karst 43, 44, 51, 93–104, 105, 149, 154, 159, 164
corrosion
  pits 129
  rates 23, 71, 88, 127, 157, 164, 245
covered karst 1, 15, 16, 91, 131, 159
cryptokarst 125, 130, 132, 135, 137, 147, 148, 150
  see also covered karst

Dabang river (Ghizhou) 106–108
deforestation 248
denudation rate 21, 23, 26, 27, 28, 55, 78, 83, 118, 131, 147
depression
  see closed depression
diagenesis 18, 242
diffuse flow 204, 228
discharge
  see flow
Disu river 141, 228, 244
dolines 53, 77, 80, 85, 101, 103, 116, 129, 147–148, 154, 159, 164, 236, 246
dolomite 14, 18–19, 54, 66, 71, 94, 121, 128, 134, 151, 153, 156, 170, 176
Dragon Bone hill (Zhoukoudian) 172
dry valleys 16, 45, 80, 84, 102, 112, 140, 146, 156, 157, 159, 226
dynamic equilibrium 103, 104, 223, 246

earthquakes 131, 169, 187
engineering, problems 37, 41, 117
epikarst 83, 84, 116, 154–155, 236
erosion surfaces 50, 95, 121, 125, 134, 159, 162, 200, 218
evaporites 11, 239
exhumed karst 148

facies, of limestones 18, 62, 98
faults 15, 20, 56, 60, 85, 93, 99, 121, 123, 124, 131, 157, 162, 169
fencong (peak cluster) 42–44, 49, 53, 64, 67, 79–87, 96, 109, 113–114, 137, 140, 143, 148, 154, 181, 237
fenglin (peak forest) 35, 42, 43, 44, 49, 64, 66–72, 78, 96, 101, 109, 113, 137, 140, 237
fissures 54, 84, 156, 162, 225
flood 77, 124, 157, 160
flow
  river 11, 12, 140, 157, 229
  springs 112, 141, 142, 159, 160, 161, 162, 167, 226, 228, 241
flowstone 187, 215, 217
fluorite 41
foraminifera 29, 150

geological controls 1–20, 242–244
geothermal water 178, 193, 202
Gigantopithecus 73
glacial
  deposits 36, 50, 184, 187
  glaciers 197, 201
  karst 52, 57, 194, 247
Goethite 216
gorges 63, 64, 104–113, 121, 124, 125, 142, 144, 183, 195, 244
gradients, river 229, 232, 234
groundwater 57, 163, 167, 168, 179, 200, 223–224, 229, 231, 232, 248
Guangxi 10, 17, 22, 49–50, 58–92, 139–143
Guilin 18, 29, 33, 36, 38–40, 58–92
Guizhou 6, 8, 17, 19, 36, 50, 51, 93–119
gypsum 11, 14, 153,156, 162, 180

haematite 216
Haihe, river 8, 12
hanging valley 107
Heshan Man 165
high altitude karst 121, 181–195
Homo erectus pekingensis (Peking Man)
  see Peking Man

Subject Index

Homo erectus yiyuanensis 165
Hong Shuihe river 140
Huangguoshu, fall (Guizhou) 105–107
Huanghe (Yellow River) 8, 12, 21, 25, 140
Huanglong valley (West Sichan) 187
Hubei and Hunnan 17, 22, 26, 49, 143–148
hydraulic gradient 87, 104, 142, 213, 219, 230–231
hydrographs 169, 204, 228, 229
hydrochemistry 186, 190, 191, 193, 194, 239, 240
hydroxyapatite 75

illite 63
infiltration rate 23, 83, 227–228
interglacials 29
internal drainage 200
inselberg 200
isotope, record 30, 150

Jiuzhaigou valley (West Sichan) 184
joints 96, 125, 129, 135, 147, 169, 175, 213
Juixian fault (Yunnan) 131

kaolinite 216
karren
  absence of 55, 159
  kamenitza (pans) 127, 212
  kluftkarren (grikes) 147, 212
  rain pits 52, 130, 211
  rillenkarren, 126, 128, 185, 211
  troughs 212
  wall solution runnels 212
karst, definitions 40, 42, 44, 62
karst plain 45, 140
karst studies, history of 32–41
karst types 15, 16, 28, 31, 45–57
karst window 229
knick point 105, 107
Kunlun Shan (Tibet) 197

lakes 146, 184, 186, 221
  Dianchi lake (Yunnan) 123
  Nam lake (Tibet) 212
  Qarhan Salt lake (Qinghai) 220
  Xi Tang lakes (Guilin) 85

landscape zones 6–10, 36, 51, 91, 109, 127, 244
landslides 89, 185
laterites 98, 124
lead 41
leisure
  see Recreational
Lijiang, river 63
Limestone Tablets 23, 203
Loess 15, 29, 32, 54–55, 153, 155–156, 174, 200, 247
Lotus Peak Biylian (Yanshuo) 69, 70
Luochuan loess sequence 29–30

Maolan Nature Preserve (Guizhou) 114
Maotiao river 94, 104, 108
mass movements 50, 172, 180
  see also landslides
micrite 66, 69–70, 149, 151, 170, 179, 203
microcaves 219
Minshan peaks (West Sichuan) 194
mixture corrosion 178
montmorillonite 217
moraine
  see glacial deposits
morphometric analysis 81, 83, 99–100, 144, 176, 207–209
mushroom rocks 128

Nanpan river (Yunnan) 134
natural bridge 107, 114, 183
neotectonics 2, 113, 116, 117, 121, 150, 155, 179, 188, 195, 244
Niangziguans, spring 159–160
notches, corrosion 77

oil and gas 41, 56, 178, 249
organic acids 203

palaeohydrology 166
palaeokarst 16, 29, 39, 54–55, 153, 159, 162, 164, 172–173, 180, 196, 201, 221
palaeomagnetic data 2, 76, 87, 171
palaeosols 29, 75, 247
pavement, limestone 52, 166, 247
peak cluster
  see Fencong

peak forest
   see Fenglin
pediments 84, 103
pediplains 103, 198
Peking Man 32, 38, 54, 151, 165, 171–174
Peneplain 171, 176, 198
   see also erosion surface
periglacial processes 200
permeability 19, 55, 179
phreatic
   caves 74, 107, 109, 111–112, 143, 213, 214, 218
   zone 65, 236
pinnacle karst 53, 125–130, 176, 177, 183, 197, 201, 206–207, 210
plain
   see karst plain
planation surface
   see erosion surface
plate tectonics 1, 2, 16, 121
polje 41, 85, 112, 118, 124, 139, 248
pollen 160
pollution 228, 241
polygonal karst 81, 102, 118, 119, 145, 207
porosity 19, 66, 232
profiles
   rivers 94, 105–106, 108–109, 157
   slopes 207

Qinling 11, 52, 154–155
Quaidam 196, 220–221
quarrying 248
Quaternary 31, 49, 99, 118, 153, 164, 176, 247
Quartz grains 215

recreational value, karst 33, 39, 74, 112, 148, 149
relict
   cave 212
   karst 91, 121, 159
resurgences
   see springs
rimstone dams 187, 188, 190
Rizhe valley (West Sichuan) 185

salt karst 220–221

scallops 73, 126
Seven Star hills (Qixing Shan) 33
Shanding Cave Man 174
Shandong 17, 53–54, 151, 163–169
Shanxi 17, 54–55, 151, 155–162
sheet flow 171, 174
Sichuan 17, 47, 52, 56, 57, 181–195
Silicates 1, 217
Sinkholes
   see swallow holes
solution rate 131
soil erosion 249
solifluction 211
sparite 18, 66, 69–70, 72, 124, 139, 149, 179, 242
speleothems 29, 149, 150, 170, 197, 217, 218
sporopollen 164, 179
springs 35, 39, 41, 53–54, 57, 84, 104, 108, 112, 117, 141, 152–153, 159, 193, 205, 226, 241, 249
   Chemao spring (Tibet) 204
   Heilongdong spring (Hebei) 229–230
   Heihu spring (Shandong) 166
   Jiala spring (Tibet) 204
   Jinan springs (Shandong) 167
   Lonzhichi spring (Shanxi) 228
   Luota springs (Guizhou) 229, 231
   Niangziguang spring (Shanxi) 159
   Queshang spring (Tibet) 204
   Wulong spring (Shanxi) 159
   Wulongtan spring (Shandong) 166
   Yudong spring (Qinling) 155
   Zhenzhu spring (Shandong) 166
Spring city (Jinan) 166
stalactites 35, 74, 113, 174, 215
stalagmites 50, 73, 74, 75, 113, 149
stone forest (Shiling) 19, 120, 124–136, 147
stylolites 151, 163
submarine spring 178
sulphates 54, 185
swallow hole 53, 80, 81, 102, 177, 197, 248
symmetry
   see Morphometric analysis

Taihang mountains 163
tafoni 219

Subject Index

tectonic controls   1–11, 13, 19–20, 47, 104, 117, 136, 149, 153, 187, 243
temperature, water   123, 155, 205, 239
terra rosa   170, 246
terrace, river   37, 63, 121, 139, 144, 169
thermoluminescence   165, 171
Tibet (Qinghai-Xizang)   6, 9, 10, 14, 20, 39, 47, 49, 52–53, 181–182, 196–219
time lag, springs   228
tourism
  see recreational
tower karst   21, 32, 33, 34, 37, 43, 44, 60, 64, 76, 118, 207
  see also fenglin
tracing, water   142
transmissivity   228, 234
travertine   183, 185, 194, 213
tritium   160, 188
tufa   84, 106, 113, 117, 131, 155, 158, 160, 183, 184, 185, 186, 189, 190, 205, 213, 217

underground streams   226, 228, 231
uranium dating   29, 50, 75, 113, 131, 149, 170, 217
uvala
  see closed depression

vadose
  caves   111, 146, 213, 214, 218
  zone   65, 83, 102, 204, 205

water resources   167, 169, 232, 233, 249
water table   76, 78, 95, 118, 125, 169
waterfall   106, 108, 113, 184
Wuyang, river (Guizhou)   107

Yiyuan Man   165
Yunnan   17, 26, 120–136

Zhoukoudian   32, 169–176
Zhujiang (Pearl river)   8, 11, 21, 24, 244
Zhuozhang river (Shanxi)   227

Springer-Verlag and the Environment

We at Springer-Verlag firmly believe that an international science publisher has a special obligation to the environment, and our corporate policies consistently reflect this conviction.

We also expect our business partners – paper mills, printers, packaging manufacturers, etc. – to commit themselves to using environmentally friendly materials and production processes.

The paper in this book is made from low- or no-chlorine pulp and is acid free, in conformance with international standards for paper permanency.

Printing: Saladruck, Berlin
Binding: Buchbinderei Lüderitz & Bauer, Berlin